N. H. McClamroch

State Models
of Dynamic Systems

A Case Study Approach

Springer-Verlag
New York Heidelberg Berlin

N. H. McClamroch

Computer, Information and
  Control Engineering, and
  Aerospace Engineering
The University of Michigan
Ann Arbor, MI 48109
USA

AMS Classification (1980): 93-01, 93A10, 39B05

With 97 Figures.

Library of Congress Cataloging in Publication Data

McClamroch, N. Harris.
  State models of dynamic systems.

  Bibliography: p.
  1. System analysis. I. Title.
QA402.M22   629.8'312   80-11570

© 1980 by Springer-Verlag New York Inc.

Printed in the United States of America.

9 8 7 6 5 4 3 2 1

ISBN 0-387-90490-5   Springer-Verlag New York
ISBN 3-540-90490-5   Springer-Verlag Berlin Heidelberg

Preface

The purpose of this book is to expose undergraduate students to the use of applied mathematics and physical argument as a basis for developing an understanding of the response characteristics, from a systems viewpoint, of a broad class of dynamic physical processes.

This book was developed for use in the course ECE 355, Dynamic Systems and Modeling, in the Department of Electrical and Computer Engineering at the University of Michigan, Ann Arbor. The course ECE 355 has been elected primarily by junior and senior level students in computer engineering or in electrical engineering. Occasionally a student from outside these two programs elected the course. Thus the book is written with this class of students in mind. It is assumed that the reader has previous background in mathematics through calculus, differential equations, and Laplace transforms, in elementary physics, and in elementary mechanics and circuits. Although these prerequisites indicate the orientation of the material, the book should be accessible and of interest to students with a much wider spectrum of experience in applied mathematical topics.

The subject matter of the book can be considered to form an introduction to the theory of mathematical systems presented from a modern, as opposed to a classical, point of view. A number of physical processes are examined where the underlying systems concepts can be clearly seen and grasped. The organization of the book around case study examples has evolved as a consequence of student suggestions. Although my personal inclination has been to embed the systems concepts in an abstract framework, the majority of undergraduate students seem best able to grasp the important ideas when they are presented through reasonably concrete illustrations. Even though no attempt is made to present a general theory it is hoped that many of the important systems concepts (model, state, time invariance, approximation, simulation, feedback, stability, etc.) are made meaningful by this approach. The reader who completes the study of the book should have a basic appreciation for the systems point of view; he should also be adequately prepared for more advanced work in systems theory, especially in the areas of control using state feedback, state estimation, optimal control, analog and digital simulation, etc.

The book actually presents a somewhat narrow view of dynamic state models, since emphasis is on state models which can be expressed in terms of ordinary differential equations. The advantage of this approach is that most upper level students in electrical or computer engineering or applied mathematics are adequately prepared, from their training and background, to handle and appreciate such models. A disadvantage in this approach is that it does not indicate the generality of many systems concepts; many other interesting state models, e.g., stochastic models, distributed models, finite state models, which have wide applications in electrical engineering, mechanics, operations research, computer science are briefly considered. On balance, however, the emphasis on state models based on ordinary differential equations is deemed justified by the author.

In order to develop competence in mathematical modeling, in mathematical analysis, or in computer simulation it is necessary to gain experience. The problem exercises at the end of each chapter can serve as a basis for obtaining such experience. The problems range from those which are easy to some which are difficult. In some of the problems answers to the stated questions can be determined analytically; in other cases only qualitative or approximate answers can be given; in some problems a complete answer to the stated questions may require a numerical simulation. No explicit indication is given in any of the problems to indicate the difficulty or the method of attack; such recognition is an important part of the learning experience. Some of the problems are modifications of material discussed in the text; other problems are not based on text material and hence include a brief statement of the relevant assumptions. It is suggested that some mathematical analysis be carried out for each individual exercise and the desired results be developed from first principles. Direct substitution into formulas developed in the text is discouraged. Although such an approach may be painful initially, the learning experience is definitely enhanced.

The nature of the material in the book is such that, if the reader is new to such an approach, considerable difficulty may be expected. The book cannot be read and comprehended either quickly or easily. The author has found that the effort required to master the material is considerable, at least for most individuals. The most common source of difficulty is the synthesis of physical argument and purpose with mathematical formalism; the development of this attitude of synthesis is the primary objective of the book. Another source of difficulty arises from the details of the mathematical calculations carried out in the individual case studies. The author has attempted to include enough description of the procedure, together with suitable equations, so that the development should be clear; however, there are numerous cases where "intermediate" calculations are not explicitly indicated. The serious reader should attempt to supply any missing details. Another source of possible difficulty relates to the use of symbolic notation throughout the notes. The author has attempted to use standard notation where possible; however, the diversity of examples considered requires repeated use of the same symbol in different case studies. There may be some confusion in the notation used for function values since function arguments are often suppressed; the distinction between function values and constants is, hopefully, clear from the context. Although the difficulty in mastering the material may be considerable for many, the end result is arrival at an understanding of the essence of the modern scientific and applied mathematics methodology.

The course ECE 355 has been taught regularly by the author since 1973 when the course was originated. This book is the consequence of repeated and earnest requests from past students who persevered with only the occasional two page handwritten handout. Appreciation is due to past students in ECE 355 and past teaching assistants in the course, C. White, J. Blight, G. Iancelescu, A. Mohdulla, D. Hegg, W. Hortos, and N. Boustany, for help in choosing illustrative problems and in refining the organization of the approach; without that motivation this book would not now be a reality.

It is important to make an additional personal acknowledgment to Professor L. G. Clark of the University of Texas, Austin. It was under his guidance that the author was introduced to the modeling process, during the years 1963–1967, through the domain of applied mechanics. Although the material considered in this text is somewhat broader in its scope, the blend of physical insight and careful mathematical technique was first demonstrated to me by Professor Clark. My debt to him for this viewpoint is considerable. Finally, I would like to acknowledge the important contributions of my family to my own educational endeavors; their support over the years is a critical factor in the existence of this book.

January, 1980   N. H. McClamroch

Contents

Chapter I

Modeling Dynamic Systems

Modeling Philosophy

Many of the physical and technological processes that are of importance in our society are difficult to describe in terms of their behavior. The cause and effect relationships are not easily discernable; there are many important variables which may play a role in the dynamic characterization of the process; the interactions between these variables may be critical. In short, many physical processes whose dynamic characterization need to be well understood are sufficiently complex that a formal approach must be followed. This formalism is best expressed in the language of mathematics. With few exceptions, our intuition, without some imposed formalism, is not capable of making sense out of the response characteristics of complex physical processes. Our objective here is to present a formal mathematical approach that is extremely useful in gaining such an understanding for a whole range of physical processes. This mathematical approach is emphasized throughout, but it is important to keep in mind that it is a tool that may be helpful in the basic task of understanding a particular physical process. Thus, the mathematical approach is by no means to be divorced from the physical process to which it is applied.

One of the difficulties here is to give a proper definition of terms. Rather than attempt to make artificial definitions it seems more suitable to proceed by assuming that the terminology used is within the reader's past experience.

Our interest, as stated above, is to set up a formal approach by which the dynamic characterization of physical processes can be understood. Physical processes may be characterized by existing hardware, conceptual devices, existing or conceptual experiments, or by biological or organizational entities. The physical variables used in describing such physical processes are typified by length, mass, temperature, neutron density, number of infectives, gross national product, etc. Without regard to the nature of the physical process, it is usually possible to recognize certain of the physical variables as having the abstract property that they represent either a local input or a local output of some subsystem, or that the physical variables are useful in describing the internal state of a subsystem. It is also typical that a physical process consists of interactions between these variables, i.e., that the physical system of interest can be represented by an interconnection of subsystems. In fact, abstract systems theory is concerned with the interconnection of subsystems and in characterizing the dynamic responses of the system in terms of the dynamic responses of its parts, i.e., the subsystems. From the physical side, this amounts to characterizing a complex device in terms of properties of the components from which the device is constructed. Thus systems theory is concerned with the total operation of the system; the subsystems are not of interest in themselves, but only as they define a part of the whole system. Of course, this all depends on the context. In one case what is considered a system may in another situation be considered only a subsystem in a more complex interconnection. This hierarchy of systems within systems presents no real difficulty since the formal mathematical approach considered is applicable at any level. Besides, our maintenance of a close connection between a physical process and an abstract system will aid in understanding both features.

The most important step in the use of systems theory as applied to a particular physical process is the translation of our understanding of the physical process into the mathematical language of systems theory. This step is often neglected in discussions of abstract system theory, but actually it constitutes the most important, and usually the most difficult, step in the whole procedure. The particular features and aspects of the physical process of interest must be recognized; the important physical variables must be recognized. Assumptions must be made about the important factors in characterizing the dynamic responses of the physical process and the factors that can be ignored. These assumptions can be formalized to obtain mathematical relationships between the physical variables. In many cases the mathematical relationships may be based on the use of standard physical theories, e.g., Newton's laws of motion, Lagrange's equations, Kirchoff's laws, etc. In some cases the mathematical relationships must be obtained on the basis of experiment; in some cases the mathematical relationships may be pure conjecture. In any event, it is usual to obtain a set of mathematical equations which describe the important physical variables of interest. It is then convenient to place these mathematical equations in a standard or normal form; the important point here is to decide on variables, called the state variables, which are *essential* in characterizing the physical process. Translating the equations in the physical variables into equations in the state variables in the standard form is then possible, the latter equations being called the state equations for the particular physical process. State equations are considered in considerably more detail shortly. In conclusion, the modeling procedure consists in the definition of a physical process and the translation of that definition into a set of descriptive state equations for that physical process.

It then remains to make use of the state model for the physical process as a tool for understanding the dynamic responses of the physical process. In essence, the state model is actually a set of instructions, often in implicit form, for generating responses under specified experimental conditions. Sometimes these model instructions can be carried out using analytical methods; often the model instructions can be carried out only by using a computer; this latter process is computer simulation. It is important to keep in mind the distinction between the physical process and its mathematical model; the model acts as a surrogate for the physical process in that mathematical experiments are performed using this model that, hopefully, mimic experiments that could be performed on the physical process. The results of this mathematical experimentation, the model responses, must eventually be interpreted as responses of the physical process. The mathematical responses are indicative of the physical responses only to the extent that the model is an accurate surrogate for the physical process.

This overall process of developing and using a mathematical model is illustrated in Figure 1-0.

The process illustrated in Figure 1-0 can be summarized as follows. A mathematical model is developed by making assumptions about the physical process. Based on this mathematical model, mathematical analysis and computer simulation can be used to determine the model responses. These model responses can finally be used to draw inferences about the responses of the physical process. The above process has been described in a sequential form; however, in practice, information and insight obtained at one stage is often used to modify what has been done at a previous stage. Hence there are often several cycles in a study, each one refining and modifying the results obtained in the previous cycle. It is also important to recognize that there is no unique end result in the process described in Figure 1-0; there are many different assumptions that might reasonably be made about any physical process and there are many different ways of implementing those assumptions in each stage of the process.

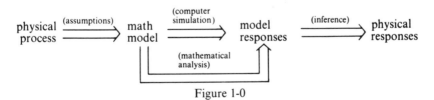

Figure 1-0

It is important to emphasize the role of assumptions in developing and using a mathematical model for a physical process. The meaning and significance of the assumptions should always be kept in mind since they form the base upon which application to the physical process is made. The assumptions should be considered to include the specification of the context in which the model is to be used. This is sometimes referred to as specification of the experimental frame of the model. The results of a study of a physical process, based on development and use of a mathematical model, are ultimately limited in their application solely by the correctness of the assumptions made about the physical process.

The objective of this text is to develop a facility for the overall modeling approach, as indicated in Figure 1-0. The emphasis in the succeeding chapters is on the formulation of mathematical models and the use of those models in determining responses, using both analytical and simulation techniques. The ideas are presented, primarily, using a number of examples or case studies. For obvious reasons, only an informal description of the "conceptual" physical process can be given; in each case an attempt is made to state any results in a context relevant to the physical process. A more specific objective of the text is to provide an introduction to systems or state concepts, as mentioned earlier. Some of these concepts are indicated by the terms: model, state, approximation, structure, simulation, feedback, for example. On a more fundamental level, an objective of the text is to demonstrate an approach to the study of dynamic physical processes which integrates physical principles used in describing a physical process with mathematical and computer simulation techniques used in characterizing the responses of that physical process. As mentioned, the approach taken in the text is to illustrate most of the important ideas in the context of a specific example or case study. A useful by-product of this approach is that some contact is made with problems in circuits, mechanics, electromechanics, chemical kinetics, fluid and thermal mechanics, physiology, ecology, etc. Knowledge of specific examples in such selected areas is clearly of importance in itself.

Some Physical Principles

Since much of the discussion of systems theory will be directly related to particular physical processes it is imperative to consider, very briefly, some relevant physical principles. The aim here is not to explain the principles, but rather to review them and to introduce some of the usual relationships between physical variables. No attempt is made to be complete; in later discussion the principles indicated here will be used in the development of concrete models. Of course, adequate familiarity with the relevant physical principles is required to obtain a good system model for any physical process.

A number of physical processes will be studied from a diverse range of scientific disciplines. Many of these physical processes are typical of those of interest to engineers, scientists, and applied mathematicians. In fact one of the advantages of the systems approach is that it is applicable to a wide range of physical processes.

A brief review of some of the standard physical principles is now given.

1. Electrical LRC Circuits

One important class of physical processes are those which can be described by electrical networks consisting of resistors, capacitors and inductors. Such devices are usually connected together in a network configuration using the standard symbols shown in Figure 1-1.

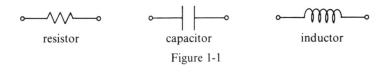

resistor   capacitor   inductor

Figure 1-1

The important physical variables are current, voltage and charge. In an ideal linear resistor the voltage drop V across a resistor is related to the current I through the resistor by

$$V = RI,$$

where R is a constant, called the resistance. In an ideal linear capacitor the charge Q on a capacitor is related to the voltage drop V across the capacitor by

$$Q = CV,$$

where C is a constant, called the capacitance. In an ideal linear inductor the voltage drop V across the inductor is related to the rate of change of current through the inductor dI/dt by

$$V = L\frac{dI}{dt},$$

where L is a constant, called the inductance. In any branch the current and charge are related by

$$I = \frac{dQ}{dt}.$$

Since resistors, capacitors and inductors are typically connected in a network there are necessarily imposed relations between the physical variables; these relations are referred to as Kirchoff's laws: the voltage drop in any loop of a network is necessarily zero, the net current into any node of the network is necessarily zero. It is important to keep in mind that there are signs associated with voltage, current, and charge depending on the convention used. In some cases an electrical network includes voltage sources and current sources. The above relations form the basis for elementary electrical network analysis. Of course there are situations when a resistor, capacitor or inductor may not be ideal and hence described as above; the particular relations between current, voltage and charge may be more complicated in the non-ideal case than the relations indicated above. More extensive discussion of electrical networks is given in the texts by Desoer and Kuh, by Smith and by Boylestad and Nashelsky.

2. Particle Dynamics

Another important class of physical processes are those which can be described in terms of the motion of ideal particles. The important physical variables for each particle are its position, as measured from a fixed reference, its velocity and its acceleration. Also of importance is the force acting on the particle. The basic relation which characterizes the motion of a particle is Newton's law: the force F on a particle and the acceleration A of the particle are related by

$$F = mA,$$

where the mass m of the particle is assumed constant. The acceleration of the particle is related to its velocity V by

$$A = \frac{dV}{dt},$$

and the velocity of the particle is related to its position H by

$$V = \frac{dH}{dt}.$$

The force, acceleration, velocity and position can be considered as "vector" variables in the usual sense. In many situations it is desired to consider the motion of a collection of particles; it is often convenient to examine free body diagrams which are used to indicate the various forces on each particle in the collection. The motion of rigid bodies, consisting of an infinity of such particles, is often of interest. The relevant equations for a rigid body can be obtained by using the above relations and integrating over the body. The resulting motion is described by equations as above in terms of the center of gravity of the body; the rotational motion of the body is described in terms of the angular position of the body as measured from a fixed reference, its angular velocity, and its angular acceleration; the torque acting on the rigid body is also important. It can be shown that the torque T acting on a rigid body and the angular acceleration α of the rigid body are related by

$$T = I\alpha,$$

where the moment of inertia I of the rigid body is assumed constant. The angular acceleration of the rigid body is related to its angular velocity ω by

$$\alpha = \frac{d\omega}{dt}$$

and the angular velocity of the rigid body is related to its angular position θ by

$$\omega = \frac{d\theta}{dt}.$$

Again the torque, angular acceleration, angular velocity and angular position can be considered as "vector" variables. The above relationships are essential ingredients in describing the motion of particles and rigid bodies; there are, however, alternative formulations of the principles based on use of Lagrange's equations which are sometimes convenient. The above is a very brief description of the basic elements in Newtonian mechanics. The principles of conservation of momentum and conservation of energy often lead to useful relationships. More extensive discussion of the principles of mechanics may be found in texts by Greenwood, Cannon, Halliday and Resnick, and Sears and Zermansky.

3. Fluid Mechanics

Some physical processes involve the motion of liquids and gases; our interest will focus on macroscopic motions rather than on the microscopic motions of molecules. The two basic relationships which govern the macroscopic motions of fluids are conservation of mass and conservation of energy. It is usual to assume that liquids are incompressible so that the volume of a liquid is a constant. Gases are usually considered to be compressible; the ideal gas law is that the pressure of a gas multiplied by its volume, divided by its absolute temperature, is always constant. Other particular principles about the motion of fluids will be introduced as required. A detailed discussion of the motion of fluids can be found in texts by Cannon, Sears and Zermansky, and Franks.

4. Thermal Dynamics

Another important class of physical processes are those which involve the transfer of heat. The basic principle of thermal dynamics is the principle of conservation of energy. Heat can be transferred between two bodies in such a way that the heat transfer rate Q to a body is related to the rate of change of the temperature of the body dT/dt by the relation

$$mC_p \frac{dT}{dt} = Q,$$

where m is the mass of the body and C_p is the specific heat of the body. The heat transfer rate depends on a number of factors. If the heat transfer depends on convection then the heat transfer rate is directly proportional to the temperature difference between the body and its surroundings. As indicated previously the temperature of a gas also depends on its volume and pressure. Detailed discussion of some of the physical principles associated with thermal dynamics can be found in texts by Adams and Rogers, and Franks.

5. Chemical Dynamics

Another class of interesting physical processes are those which are characterized by chemical reactions. The rate at which a chemical reaction occurs may be quite complicated since the rate generally depends on the amount of reactants as well as the particular nature of the reactions. The general principles of conservation of mass and conservation of energy are often useful as a means for describing certain aspects of the reactions. Information about chemical reaction rates can be found in texts by Walas, Frost and Pearson, and Franks.

6. Other

Only brief descriptions of several areas where standard physical principles are available have been described. There are numerous other scientific areas where physical principles are well established and accepted. Such areas include electromechanical energy conservation, nuclear reactions, operations research, physiology, population biology, economics. Certain principles have even been suggested in the fields of medicine, sociology, linguistics, anthropology, and history which might serve as a basis for the use of the systems approach. Again, it should be emphasized that correct understanding and application of physical principles or theories usually requires a great deal of knowledge about the particular field. A survey of some of these mentioned fields, where some attention is given to the development of physical principles, is available in the references.

The above discussion should not be interpreted as implying that there are always well-developed physical principles that always lead directly to the formulation of useful mathematical models for the physical process. Sometimes this may be the case, but often formulation of a systems model depends on a certain amount of conjecture and experimentation. Experimental work may be required simply to determine the values of certain parameters or to determine the functional form of certain more basic relationships. Experimental work is also valuable as a means for determining the important factors to be considered in the formulation of the systems model. Finally, it is important to mention that all physical variables generally have some physical units. As previously, it is traditional to write equations where the physical units are not indicated explicitly. But it is essential to keep in mind that any equation denotes equality of the associated numerical values as well as equality of the associated physical units. In some cases it is possible to locate subtle mistakes by simply checking the correctness of the units in an equation. There are many different measurement systems which can be used, e.g., the English or metric system. The particular set of units used is not of particular importance so long as the physical units are used in a consistent manner. As is usual, no physical units are indicated in the later discussion except in the case where numerical data from analysis or simulation is indicated. Throughout, it is assumed that physical variables can be either positive or negative unless specified otherwise. Physical constants are always assumed to be positive.

This text cannot be considered a substitute for detailed work in the general physical areas briefly discussed in this section. No attempt is made here to give a complete development of the basic principles of circuit theory, mechanics, etc. Complete texts which do develop the fundamental principles in the various physical areas are available, as indicated in the list of references.

State Models

Based on the physical principles relevant to a particular physical process the equations which characterize the physical process can usually be placed in a state model or systems model. It is usually possible to decide upon the input and output variables of interest; this subjective decision depends on what aspects of the dynamic responses of the physical process are of interest. The input and output variables are typically certain of the physical variables, the input variable relates the effects of the "external world" on the system while the output variable relates the effects of the system on the "external world." Next, state variables for the system should be identified. These variables should have the following property: *If the current state is known and the future input is known then the future state and output are uniquely determined.* That is, the state variables of a system completely characterize the past of the system, since the past input is not required to determine the future output of the system. This seemingly elementary notion of the state, defined implicitly by this notion, is quite important. In fact, the modern approach to dynamic systems is fundamentally based upon this concept of state. Although the state notion has been explicitly emphasized only during the last several decades such notions have been recognized for centuries. For example, the 18th century mathematician Laplace recognized that

> we ought then to regard the state of the universe as the effect of its antecedent state and the cause of the state that is to follow.

This concept of the state of a system forms the basis for the modeling approach developed in this text. Unfortunately, it is not always clear how to choose the state variables; they may be certain of the physical variables but they need not be; further, there is no unique choice of state variables. Although no general prescription can be given for choosing state variables general guidelines can be given, e.g., in electrical LRC circuits the charge on each capacitor and the current through each inductor in the network usually serve to define the state of the network; in a mechanical system the position and velocity of each particle, or of each rigid body, usually serve to define the state of a mechanical system. In many other physical processes the choice of state variables may be rather direct, based on experience. In some cases, the choice of state variables is extremely difficult. Once state variables are chosen the mathematical equations characterizing the state variables, the state equations, can be derived.

Depending on the particular form of the physical equations used to describe the physical process the state equations may be in one of many mathematical forms. It is possible to classify state equations on the basis of their mathematical structure. Time is usually an independent variable in a state model. Sometimes the time variable is considered to be a discrete variable in which case the state model is typically described by recursive equations; in other cases the independent variable time is considered to be real valued. Sometimes there can be additional independent variables in which case the state model is said to be distributed; such state models can often be described by partial differential equations. If time is the only independent variable then the state model is said to be lumped. Further, the state model may include random effects in which case the state model is said to be stochastic. If no such effects are included the state model is said to be deterministic. In this book attention is focused on continuous time, lumped, deterministic state models for physical processes.

To be more explicit, the state equations considered in this text are assumed to be described by ordinary differential equations. In particular, if u denotes the input variable and y denotes the output variable it is assumed that the rate of change of each state variable at time t depends on the values of the state variables at time t and the value of the input variable at time t according to

$$\frac{dx_1}{dt} = f_1(x_1, \ldots, x_n, u),$$

$$\vdots$$

$$\frac{dx_n}{dt} = f_n(x_1, \ldots, x_n, u).$$

Further, the output variable is assumed to depend on the state variables and the input variable according to an algebraic equation

$$y = g(x_1, \ldots, x_n, u).$$

For generality, the model is expressed in terms of n state variables labeled $x_1, \ldots, x_n$.

It is claimed that if a mathematical model for a physical process can be written in the above form then in fact the variables $x_1, \ldots, x_n$ do have the state property mentioned earlier. The above state equations will serve as the basis for the application of systems theory to physical processes; the state equations, in the above first order form, are said to be in the standard or normal form. Only a single input and a single output variable are indicated; additional input and output variables could be included with little added conceptual difficulty.

The state equations above are written so that the state model is defined by the integer $n \geqslant 1$ called the order of the state equations and the functions $f_1(\ldots), \ldots, f_n(\ldots)$ and $g(\ldots)$. The state model is a systems model valid for any initial state and any input function; this fact is particularly important since a major interest is in characterizing the dependence of the output response of the system on the initial state of the system and on the input to the system. The state equations in the above form also serve as a very general model for many physical processes; that is, it is often possible to develop state models of the above form for many physical processes; thus there is a large range of potential application of any mathematical systems theory based on the above state equations. Finally, the above state equations are in a form which is particularly convenient for mathematical analysis as well as for computer simulation. Although the previous discussion of modeling systems and the notion of state is extremely general our attention will be focused primarily on the use of these concepts in the particular case where the state equations can be placed in the above differential form.

This is the appropriate point to mention that the previously defined differential state model, defined in terms of arbitrary functions $f_1, \ldots, f_n$ and g, is sufficiently general to include models developed for many physical processes from essentially all areas of science. However, this extreme generality means that very few specific statements can be made about the nature of all such models. A special class of state models, of much practical importance in applications, are those for which the functions $f_1, \ldots, f_n$ and g have a linear mathematical structure so that the state equations can be written as

$$\frac{dx_1}{dt} = a_{11}x_1 + \cdots + a_{1n}x_n + b_1 u,$$

$$\vdots$$

$$\frac{dx_n}{dt} = a_{n1}x_1 + \cdots + a_{nn}x_n + b_n u,$$

$$y = c_1 x_1 + \cdots + c_n x_n + d u,$$

where $a_{11}, \ldots, a_{nn}, b_1, \ldots, b_n, c_1, \ldots, c_n, d$ are constants. State models with such a special structure are said to be *linear* state models. This class of linear state models includes many, but by no means all, of the models of interest in applications. The advantage in developing and using such a linear state model is that the particular mathematical structure of the state equations allows use of powerful analytical methods. In particular the use of Laplace transformation methods (reviewed in Appendix B) is an extremely powerful mathematical tool for analyzing such models. Hence in the following chapters the (mathematically easier) models with linear structure are studied in Chapters II, III, IV; the (mathematically more difficult) models without such a linear structure are studied in Chapters V, VI, VII. Differential models with even more complicated features are studied in Chapter VIII; in Chapter IX state models which are discrete in some sense are examined.

Some Systems Concepts

In this section a number of useful systems concepts are introduced. All of the discussed concepts can be generalized but it is sufficient to present them here in terms of the state model

$$\frac{dx_1}{dt} = f_1(x_1, \ldots, x_n, u),$$

$$\vdots$$

$$\frac{dx_n}{dt} = f_n(x_1, \ldots, x_n, u),$$

$$y = g(x_1, \ldots, x_n, u).$$

As indicated the state variables are not unique; there are generally many different but mathematically equivalent ways of choosing state variables for a given physical process. It is important to recognize the relationship between different state models for the same physical process. First, suppose that the variable change

$$x_i = z_i + d_i, \quad i = 1, \ldots, n$$

is made, where d_i, $i = 1, \ldots, n$ are constants. Then if $x_1, \ldots, x_n$ are state variables for a physical process satisfying the above state equations it is easy to show that $z_1, \ldots, z_n$ can also be used as state variables for the same physical process, satisfying

$$\frac{dz_1}{dt} = f_1(z_1 + d_1, \ldots, z_n + d_n, u),$$

$$\vdots$$

$$\frac{dz_n}{dt} = f_n(z_1 + d_1, \ldots, z_n + d_n, u),$$

$$y = g(z_1 + d_1, \ldots, z_n + d_n, u).$$

Such state models are related by a *translation transformation*.

Now consider the variable change

$$x_i = \sum_{j=1}^{n} p_{ij} z_j, \quad i = 1, \ldots, n$$

where the determinant of the matrix $[p_{ij}]$ is assumed to be nonzero so that it is possible to write

$$z_i = \sum_{j=1}^{n} q_{ij} x_j, \quad i = 1, \ldots, n.$$

Then if $x_1, \ldots, x_n$ are state variables for a physical process satisfying the above state equations it is easy to show that $z_1, \ldots, z_n$ can also be used as state variables for the same physical process, satisfying

$$\frac{dz_1}{dt} = \sum_{j=1}^{n} q_{ij} f_j \left(\sum_{j=1}^{n} p_{1j} z_j, \ldots, \sum_{j=1}^{n} p_{nj} z_j, u \right),$$

$$\vdots$$

$$\frac{dz_n}{dt} = \sum_{j=1}^{n} q_{nj} f_j \left(\sum_{j=1}^{n} p_{1j} z_j, \ldots, \sum_{j=1}^{n} p_{nj} z_j, u \right),$$

$$y = g \left(\sum_{j=1}^{n} p_{1j} z_j, \ldots, \sum_{j=1}^{n} p_{nj} z_j, u \right).$$

The two sets of state variables are said to be related by a *linear transformation*. If two sets of state variables are related by either a translation or a linear transformation then the state variables are said to be *equivalent* since either set can be used as a state model for the physical process.

In the notion of state equivalence the number of state variables in the two equivalent sets is necessarily the same. However, it is possible to develop different sets of state equations, containing different numbers of state variables, for the same physical process. Such state variables are said to be *input-output equivalent*. Clearly if two sets of state variables are state equivalent then they are input-output equivalent. One of the most important problems in the formulation of a state model of a physical process is the derivation of a state model with the smallest possible number of state variables; such a state model is said to be a minimal state model for the system.

These notions of equivalence are important since any of a number of different state models can be used as a mathematical description of a physical process. It is usual to use a state model which gives some particular insight into the physical process or which leads to relatively easy computations in analysis; minimal state models are important precisely for this reason.

Our interest is in both the formulation of state models and also the analysis of the state models as a way of understanding the dynamic responses. Thus it is important to discuss, briefly, the notion of a solution to the state equations. With suitable assumptions about the functions

$f_1, \ldots, f_n$, g, it is possible to show the following: *For given initial state values $x_1(0), \ldots, x_n(0)$ and a given input function $u(t)$ defined for $t \geq 0$ there exists a unique solution of the state equations $x_1(t), \ldots, x_n(t)$ defined for $t \geq 0$* i.e., the functions $x_1(t), \ldots, x_n(t)$ satisfy the state equations exactly for all $t \geq 0$, and hence there exists a unique output function $y(t)$ defined for $t \geq 0$. In fact this solution property is simply a restatement of the state property mentioned earlier; in order to determine the future output the current state and the future input are required. The initial or starting time has been assumed to be $t = 0$; this is simply a convenience. Further the notation $x_1(t), \ldots, x_n(t)$ has been used to denote a particular solution of the state equations; this notation has some logical difficulties but it is simple and will be used throughout.

It is sometimes convenient to use the terminology of the *zero input response*. This is the output response of the system, according to the state equations, assuming that the input function is zero. Of course, the zero input response depends on the initial state of the system. The terminology of the *zero state response* refers to the output response of the system assuming that the initial state is zero. Of course, the zero state response depends on the input function. The *general response* of the system depends both on the initial state and the input function. Although it is not always true, there is an important class of systems, namely linear systems, for which the general response is the sum of the zero input response and the zero state response.

It is often possible to obtain a good understanding of the response properties of a system by considering the response to particular input functions. Some of the typical input functions often considered are constant function, a polynomial function, or a sinusoidal function. Of course, other more complicated input functions may also be of interest, depending on the nature of the particular physical process.

A particularly important class of state responses of the state equations are those which are *periodic* or oscillatory. If for some initial state and some input function the solution of the state equations satisfy

$$x_i(t + T) = x_i(t), \quad i = 1, \ldots, n$$

for all $t \geq 0$, where $T \geq 0$, then the state response is said to be periodic. The smallest such value of T is said to be the period of the oscillation. A special class of periodic state responses are those which are constant functions, i.e., for some initial state and some input function the solution of the state equations satisfy

$$x_i(t) = x_i(0), \quad i = 1, \ldots, n$$

for all $t \geq 0$. Such constant state responses are said to define an *equilibrium state*, denoted by $x_1(0), \ldots, x_n(0)$.

An important property of a class of responses of the state equations is the property of *stability*. An equilibrium state is stable, roughly, if for any

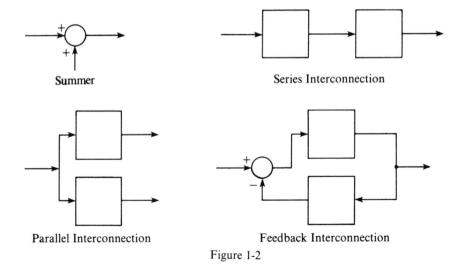

Figure 1-2

initial state close to the equilibrium state the state response tends, in the limit as time increases, to the equilibrium state. The notion of stability is examined in more detail in the later chapters.

As mentioned previously one emphasis of systems theory is the view of a system as an *interconnection of subsystems*. The state equations serve as the primary systems model but it is often convenient to give a pictorial representation of the state equations in terms of the "information flow" through the subsystems. This is most commonly done using signal flow graphs or block diagram interconnections; the latter scheme is used here. Four simple interconnections are shown in Figure 1-2.

The most elementary interconnections are those of a summer, a series connection of two subsystems, a parallel connection of two subsystems and a feedback connection of two subsystems. The arrows on the signals serve to denote the input and output of each subsystem. A complex interconnection generally consists of several combinations of the above simple interconnections. Each subsystem may be described by state equations itself; more simply the subsystem may be described by a functional relationship between its input and output. Where appropriate the subsystem may be described by the transfer function from input to output. As an example, the state equations indicated previously can be described by the interconnection shown in Figure 1-3, where the integral sign is used to indicate the integration operator.

Such a block diagram is referred to as the all-integrator block diagram corresponding to the state equations; note that the state variables $x_1, \ldots, x_n$ are the outputs of the n integrators in the interconnection. Such pictorial block diagrams are often useful as a means of displaying in an explicit way the interactions between the variables. Block diagrams can also be useful for development of analog computer simulations.

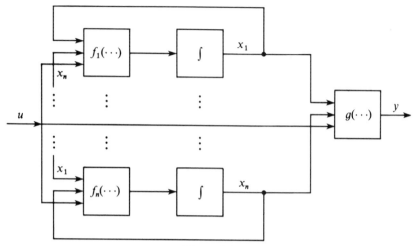

Figure 1-3

The previous discussion has given a very brief survey of some of the most important features of mathematical systems theory. Only a few of the most important topics were mentioned and they were discussed only in general terms. The objective of this discussion has been to indicate the overall focus of mathematical systems theory and to point the way for the material which follows. More extensive treatments of mathematical systems theory are available in the references, particularly the texts of Close and Frederick, Perkins and Cruz, Luenberger, and Polak and Wong.

Use of State Models

The formulation of state models for physical processes has been discussed and a few systems concepts, relevant to state models, have been introduced. It is appropriate now to mention how the systems approach can be used in terms of gaining insight into the dynamic responses of a physical process.

The most direct use of a state model is as a basis for the analysis of the dynamic responses of the physical process. This, of course, involves analysis of the response properties of the state equations for the physical process. This analysis may be carried out analytically or it may be carried out using a computer; this latter process is called computer simulation. In most cases both mathematical analysis as well as computer simulation are required and they usually yield complementary information. Again, it should be emphasized that both approaches to understanding a physical process depend on use of the state equations.

A related but usually more difficult use of state models is as a basis for the modification of the physical process in order to achieve some desired objective. This process is usually referred to as synthesis or design of the physical process. One important approach to system design involves the addition of feedback; considerably improved response properties can often be achieved; the effects of feedback are considered in more detail later. An important mathematical tool in system design is that of optimization; it is often possible to choose parameter settings or functional relationships so as to achieve some specified objective in a "best way." Optimization is a particularly powerful approach to system design, especially with accessability to large digital computers. Thus the problem of system design depends in a critical way on the use of a state model for the physical process.

The use of state models as a basis for analysis and design of physical processes is particularly common. But, state models can be useful in other ways as well. For instance, state models may be useful for identification of the value of certain physical variables or parameters, state models may give useful information regarding efficient collection of experimental physical data, state models may be used as a means for verification of assumptions or theories about the physical process. Thus, state models or system models of a physical process can be used in many important ways; in this text no special attention is given to optimization, identification, data collection, etc. but all of these topics depend on availability of a state model for a physical process.

In summary, state models are useful in developing insight into some aspects of a physical process. This insight may be qualitative in nature or it may involve quantitative information about the dynamic responses of the physical process. State models are useful in obtaining both types of information. Qualitative information is usually best obtained by a mathematical analysis of the state model; quantitative information is usually best obtained by a computer simulation of the state model. But of course both kinds of information are useful.

Chapter II

First Order Linear State Models

In this chapter physical systems which require only one state variable are examined; it is further assumed that the state equations are linear so that Laplace transforms, as discussed in Appendix B, can be used with benefit. A brief discussion of some basic systems notions for such state models is first given. Then these systems concepts are used to develop and analyze state models for several particular physical processes.

Some Systems Theory

In this section state models of the form

$$\frac{dx}{dt} = ax + bu,$$
$$y = cx$$

are examined; here u is the input variable, y is the output variable and x is a state variable. The state model is defined by the constants a, b, c. A block diagram interconnection is given in Figure 2-1 where the symbol $1/s$ is used to denote the integration operator.

Note the implicit feedback structure of the state equations which is obvious from the block diagram.

Due to the simplicity of the state equations it is a relatively easy matter to determine an explicit representation for the solution response as a function of the initial state and the input function. Suppose that $U(s)$ denotes the Laplace transform of $u(t)$, $Y(s)$ denotes the Laplace transform of $y(t)$ and $X(s)$ denotes the Laplace transform of $x(t)$. Then from the state

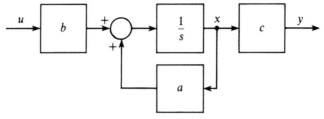

Figure 2-1

equations obtain

$$sX - x(0) = aX + bU,$$

$$Y = cX,$$

where $x(0)$ denotes the initial state. Thus

$$X = \frac{x(0)}{s - a} + \frac{bU}{s - a},$$

$$Y = \frac{cx(0)}{s - a} + \frac{cb}{s - a} U.$$

Using the notion of the convolution integral then for $t \geqslant 0$

$$x(t) = e^{at}x(0) + \int_0^t e^{a(t-\tau)}bu(\tau) \, d\tau$$

and

$$y(t) = ce^{at}x(0) + \int_0^t ce^{a(t-\tau)}bu(\tau) \, d\tau$$

which are explicit expressions for the state and output responses as a function of the initial state and the input function. The first term in the above expression is the zero input response while the second term is the zero state response. Thus in this case the general response is the sum of the zero input response and the zero state response.

It is convenient to introduce some terminology. The expression

$$G(s) = \frac{cb}{s - a}$$

is called the transfer function of the system; the transform of the zero state output response is the product of the transfer function and the transform of the input function. The expression

$$g(t) = ce^{at}b$$

is called the weighting or impulse response function of the system; note that the transfer function is the transform of the weighting function. The denominator polynomial of the transfer function

$$d(s) = s - a$$

is called the characteristic polynomial. If $a < 0$, i.e., the zero of the characteristic polynomial is negative, then the zero input response always tends to zero as $t \to \infty$, for any initial state; the state equations are said to be stable. If $a > 0$ the state equations are said to be unstable.

It is clear that the state equation indicated, with one state variable, is a minimal realization for the system assuming that $cb \neq 0$. Any state model for the same system with more than one state variable would certainly not be minimal.

Now consider the response of the system to certain standard input functions assuming that $a < 0$, i.e., the system is stable.

Consider constant input $u(t) = \bar{u}$ for $t \geqslant 0$. If $dx/dt = 0 = ax + b\bar{u}$, then $x = -b\bar{u}/a$ is an equilibrium state. Thus, if $x(0) = -b\bar{u}/a$ then $x(t) = -b\bar{u}/a$ and $y(t) = -cb\bar{u}/a$ for all $t \geqslant 0$. More generally for any initial state the state equations can be solved, using Laplace transforms, to obtain

$$y(t) = \left(cx(0) + \frac{cb\bar{u}}{a} \right) e^{at} - \frac{cb\bar{u}}{a}$$

so that

$$y(t) \to - \frac{cb\bar{u}}{a} \quad \text{as } t \to \infty.$$

This result, assuming $x(0) = 0$, could also be obtained using the transfer function since

$$Y = \left(\frac{cb}{s-a} \right) \left(\frac{\bar{u}}{s} \right).$$

Using the final value theorem (see Appendix B) it follows that

$$\lim_{t \to \infty} y(t) = \lim_{s \to 0} sY = - \frac{cb\bar{u}}{a}.$$

Now consider a sinusoidal input function of the form $u(t) = \bar{u} \cos \omega t$ for $t \geqslant 0$ for constant $\bar{u}$. The transform of the output response is

$$Y = \frac{cx(0)}{s-a} + \frac{cb\bar{u}s}{(s-a)(s^2 + \omega^2)}.$$

After a partial fraction expansion obtain

$$Y = \left[cx(0) + \frac{cba\bar{u}}{\omega^2 + a^2} \right] \left(\frac{1}{s-a} \right)$$

$$+ \left(\frac{cb\bar{u}}{\omega^2 + a^2} \right) \left(\frac{-as + \omega^2}{s^2 + \omega^2} \right)$$

so that

$$y(t) = \left[-cx(0) + \frac{cba\bar{u}}{\omega^2 + a^2} \right] e^{at}$$

$$+ \frac{cb\bar{u}}{\sqrt{\omega^2 + a^2}} \cos \left[\omega t + \tan^{-1} \left(\frac{-\omega}{-a} \right) \right].$$

If $a < 0$ so that the system is stable then

$$y(t) \rightarrow \frac{cb\bar{u}}{\sqrt{\omega^2 + a^2}} \cos\left[\omega t + \tan^{-1}\left(\frac{-\omega}{-a}\right)\right] \quad \text{as } t \rightarrow \infty.$$

This latter expression is referred to as the steady state response to a sinusoidal input function $u(t) = \bar{u}\cos\omega t$ for $t \geqslant 0$. Note that the complex function $G(j\omega)$ can be written in polar form as

$$G(j\omega) = M(\omega)e^{j\phi(\omega)},$$

where

$$M(\omega) = \frac{cb}{\sqrt{\omega^2 + a^2}},$$

$$\tan\phi(\omega) = \frac{-\omega}{-a}.$$

Thus it is easily verified that the steady state output response can be written as

$$y(t) = M(\omega)\bar{u}\cos(\omega t + \phi(\omega)).$$

In fact this is a particularly easy way to determine the steady state response to a sinusoidal input function solely from the transfer function $G(s)$. The complex function $G(j\omega)$, as a function of the frequency ω of the sinusoidal input function, is particularly important; this complex function $G(j\omega)$ is called the frequency response function. It is usual to indicate the amplitude $M(\omega)$ and phase $\phi(\omega)$ graphically as a function of the parameter ω; these plots are often given on a log-log scale and are referred to as Bode plots. These plots indicate the steady state response properties of a system to a sinusoidal input function, as a function of the frequency of the input function.

Only a very brief examination of first order linear state models has been given; many additional theoretical results could be derived. In the following sections of this chapter several physical processes are considered which can be described by first order state equations.

Case Study 2-1: Radioactive Decay of Cesium

Consider a tank containing a certain quantity of a radioactive material, e.g., the isotope Cesium 132. Such radioactive material decays, i.e., the mass of the material changes as radiation is emitted. On the basis of statistical assumptions about the radioactive decay process it is usually assumed that the rate at which the mass decays is proportional to the mass itself. Assuming that radioactive material can be added to the tank at the rate r, the change in the mass m of the radioactive material in the tank is described by

$$\frac{dm}{dt} = -Km + r,$$

where K is a constant depending on the properties of the radioactive material. It is clear that this simple first order linear differential equation is in state variable form; to be consistent with the notation used previously define the state variable $x = m$ and input variable $u = r$. The output variable y is the mass m of cesium in the tank. Thus the state model is

$$\frac{dx}{dt} = -Kx + u,$$

$$y = x.$$

Our objective in this example is to analyze the response properties of the process so that the relation between the rate at which radioactive material is added to the tank and the mass of radioactive material in the tank can be clarified.

A state description of the process has been given; a simple block diagram representation for the process is indicated in Figure 2-1-1.

Some typical response properties of the system are now determined.

In this simple case it is possible to determine an analytical expression for the general response function. Assuming the initial state is M and $u(t)$ for $t \geqslant 0$ is the input function, then the transform of the state equation leads to

$$sX - M = -KX + U,$$

where $U(s)$ and $X(s)$ are Laplace transforms of $u(t)$ and $x(t)$, so that

$$X = \frac{M}{s + K} + \frac{U}{s + K}$$

and

$$y(t) = x(t) = e^{-Kt}M + \int_0^t e^{-K(t-\tau)}u(\tau)\,d\tau.$$

Thus the impulse response or weighting function is

$$g(t) = e^{-Kt}$$

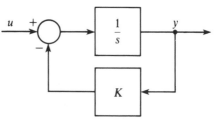

Figure 2-1-1

and the transfer function is

$$G(s) = \frac{1}{s + K} \,.$$

Since $K > 0$ the system is stable.

Now consider the response to constant input $u(t) = R$ for $t \geqslant 0$, where radioactive material is added at a constant rate. The response is thus

$$y(t) = e^{-Kt}M - \frac{R}{K}e^{-Kt} + \frac{R}{K}$$

and

$$y(t) \to \frac{R}{K} \quad \text{as } t \to \infty,$$

i.e., the mass of radioactive material in the tank tends to a constant value, independent of the initial mass M.

Case Study 2-2: Temperature in a Building

In this example the temperature in a building is considered where the outside air surrounding the building is at a generally different temperature than the building. The temperature inside the building is considered to be uniform throughout, say at temperature T. The air outside the building is considered to be at a uniform and constant temperature T_a. Heat can be generated within the building by operation of a furnace, which is assumed to supply heat uniformly to the building at the rate Q. Our interest is to relate the rate at which the furnace supplies heat to the building, considered as the input, to the temperature of the building, considered as the output. The basic physical relationship here is that the total rate of change of heat within the building, given by $mC_p(dT/dt)$ where m is mass of air in the building and C_p is the heat capacity, equals the rate at which heat is supplied by the furnace Q plus the rate at which heat is transferred to the building from the surrounding air. But the rate of heat transfer into the building from the surrounding air is proportional to the temperature difference between the building and the surrounding air, namely $K_a(T - T_a)$ for constant K_a. Thus, the rate of change of temperature is given by

$$mC_p \frac{dT}{dt} = -K_a(T - T_a) + Q$$

or

$$\frac{dT}{dt} = -\frac{1}{\tau}(T - T_a) + KQ,$$

where $\tau = K_a/mC_p$ and $K = 1/mC$.

Assuming that the surrounding air is at constant temperature T_a it is convenient to choose the temperature difference $x = T - T_a$ as the state variable; the input variable $u = Q$ and the output variable $y = T$; thus the state model is given by

$$\frac{dx}{dt} = -\frac{1}{\tau} x + Ku,$$

$$y = x + T_a.$$

Note that since $y = x + T_a$ these equations are not exactly of the form considered previously; however, with minor modifications those ideas are applicable here.

A block diagram representation, based on the above state equations, is shown in Figure 2-2-1.

It is a relatively simple matter to determine an exact analytical expression for the time response by taking the transform of the state equation. Assuming the initial state is $(T_0 - T_a)$ and the input is $u(t)$ for $t \geq 0$ then

$$sX - (T_0 - T_a) = -\frac{1}{\tau} X + KU,$$

where $U(s)$ and $X(s)$ are Laplace transforms of $u(t)$ and $x(t)$, so that

$$X = \frac{T_0 - T_a}{s + 1/\tau} + \frac{KU}{s + 1/\tau};$$

thus

$$x(t) = e^{-t/\tau}(T_0 - T_a) + \int_0^t e^{-(t-\lambda)/\tau} Ku(\lambda) \, d\lambda;$$

consequently the output response is

$$y(t) = T_a + e^{-t/\tau}(T_0 - T_a)$$
$$+ \int_0^t e^{-(t-\lambda)/\tau} Ku(\lambda) \, d\lambda.$$

Due to the presence of the constant T_a it is not possible to define an impulse response function or a transfer function with input u and output y. But it is clear that

$$g(t) = e^{-t/\tau}K$$

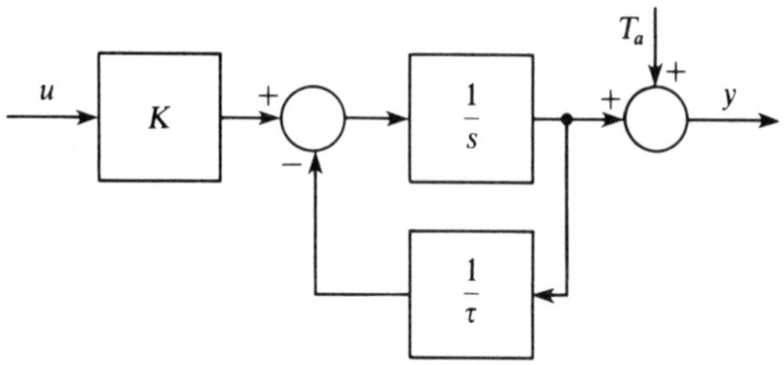

Figure 2-2-1

is the impulse response function for the related system with input u and output $y - T_a$. Similarly the transfer function from u to $y - T_a$ is

$$G(s) = \frac{K}{s + 1/\tau}.$$

The variable $y - T_a$ is obvious in the block diagram representation of Figure 2-2-1.

From the above analysis it is clear that the zero input response is

$$y(t) = T_a + e^{-t/\tau}(T_0 - T_a).$$

Thus, if no heat is supplied to the building by the furnace then the temperature of the building tends to the temperature of the surrounding air, no matter what is the initial temperature of the building. The rate at which this change occurs is characterized by the parameter τ, which is often called the time constant of the building.

Now consider the case where the furnace is in continual operation and transfers heat to the building at a constant rate so that $u(t) = \overline{Q}$ for $t \geqslant 0$. Then, from the above expression for the output response it is easily determined that

$$y(t) = T_a + e^{-t/\tau}(T_0 - T_a)$$
$$+ \tau K \overline{Q}(1 - e^{-t/\tau}).$$

Thus if heat is transferred to the building by the furnace at a constant rate the temperature of the building satisfies

$$y(t) \rightarrow T_a + \tau K \overline{Q} \quad \text{as } t \rightarrow \infty,$$

that is, the ultimate temperature of the building exceeds the outside temperature by the quantity $\tau K \overline{Q}$.

Case Study 2-3: An RC Circuit

In this example an extremely simple electrical circuit is examined which contains a single ideal resistor and a single ideal capacitor. A schematic diagram for the circuit is shown in Figure 2-3-1.

The input is the voltage e_i applied at the indicated terminals; it is assumed that the applied voltage source e_i has zero impedance. The output of the circuit is the voltage e_0 across the resistor. Let Q denote the charge

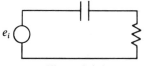

Figure 2-3-1

on the capacitor and let I denote the current through the capacitor; then the total voltage drop in the single loop is

$$\frac{Q}{C} + RI - e_i = 0,$$

where C is the capacitance and R is the resistance of the circuit. In addition the charge and current relationship is

$$\frac{dQ}{dt} = I.$$

It is natural to choose the charge Q on the capacitor as the state variable x; the input variable $u = e_i$ and the output variable $y = e_0 = RI$. Thus the state model for the circuit is given by

$$\frac{dx}{dt} = -\frac{1}{RC}x + \frac{1}{R}u,$$

$$y = -\frac{1}{C}x + u.$$

Our objective here is to relate the response properties of the output voltage to the initial state of the circuit and to the input voltage.

A simple block diagram for the process is given in Figure 2-3-2.

The response properties of the system are now examined. It is possible to determine an expression for the general response as follows. Taking the transform of the state equation, assuming Q_0 is the initial state and $u(t)$, $t \geqslant 0$ is the input, obtain

$$sX - Q_0 = -\frac{1}{RC}X + \frac{1}{R}U,$$

where $U(s)$ and $X(s)$ are Laplace transforms of $u(t)$ and $x(t)$, so that

$$X = \frac{Q_0}{s + 1/RC} + \frac{U}{R(s + 1/RC)}$$

and consequently

$$x(t) = e^{-t/RC}Q_0 + \int_0^t e^{-(t-\tau)/RC}\frac{u(\tau)}{R}\,d\tau.$$

Thus the output voltage is

$$y(t) = -\frac{Q_0}{C}e^{-t/RC} - \frac{1}{RC}\int_0^t e^{-(t-\tau)/RC}u(\tau)\,d\tau + u(t).$$

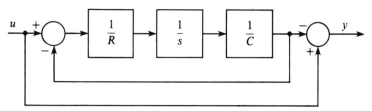

Figure 2-3-2

The transfer function for the circuit is given by

$$G(s) = \frac{s}{s + 1/RC} .$$

Since $RC > 0$ the system is stable.

Now consider the response to a sinusoidal input voltage $u(t) = \bar{e}_i \cos \omega t$ for $t \geqslant 0$. Then the output is given by

$$y(t) = -\frac{Q_0}{C} e^{-t/RC} - \frac{\bar{e}_i}{RC} \int_0^t e^{-(t-\tau)/RC} \cos \omega \tau \, d\tau + \bar{e}_i \cos \omega t.$$

The integral above can be evaluated with some difficulty. Alternatively, from the frequency response function

$$y(t) \to M(\omega) \bar{e}_i \cos(\omega t + \phi(\omega)) \quad \text{as } t \to \infty,$$

where, from the transfer function,

$$M(\omega) e^{j\phi(\omega)} = \frac{j\omega}{j\omega + 1/RC} ;$$

thus

$$M(\omega) = \frac{\omega}{\sqrt{\omega^2 + (1/RC)^2}} ,$$

$$\tan \phi(\omega) = \frac{1}{\omega RC} .$$

Thus the frequency response functions have the following properties: if $\omega = 0$, $M(\omega) = 0$ and $\phi(\omega) = \pi/2$; for $\omega > 0$, $0 < M(\omega) < 1$, $0 < \phi(\omega) < \pi/2$; as $\omega \to \infty$, $M(\omega) \to 1$ and $\phi(\omega) \to 0$. Bode plots of these frequency response functions are shown qualitatively in Figure 2-3-3. Since $\phi(\omega) > 0$ such a circuit is called a phase lead network.

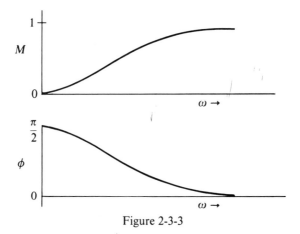

Figure 2-3-3

Exercises

2-1. Consider the radioactive decay process as described in Case Study 2-1. The half life of radioactive Cesium 132 is 9.7 days. Assuming that initially there is 1.0 kg of Cesium in a closed container, how much Cesium should be added to the container at the end of the first day so that at the end of the second day there is also 1.0 kg of Cesium in the container?

2-2. Consider the Case Study 2-2 describing the temperature in a building. The parameter values in the model are given by

$$\tau = 5.0 \text{ hr}, \quad T_a = 0°\text{F}, \quad K = 0.05°\text{F}/\text{BTU}.$$

(a) What is the temperature in the building during a 2.0 hour time period if the initial temperature in the building is 0°F and the heat transfer rate Q to the building is given by the plot below?

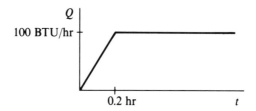

(b) What is the temperature in the building during a 2.0 hour time period if the initial temperature in the building is 65°F and the heat transfer rate Q to the building is given by the plot below?

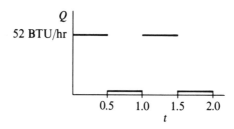

2-3. For the RC circuit described in Case Study 2-3 suppose that the capacitance and resistance of the circuit are

$$C = 0.5 \times 10^{-6} \text{ farad}, \quad R = 100.0 \text{ ohm}.$$

Determine, if it is possible, an input voltage waveform $e_i(t)$ for $t \geq 0$ so that if the initial charge on the capacitor is $Q(0) = 0.0$ coulombs the charge on the capacitor satisfies

$$Q(t) = 10^{-6} \text{ coulombs} \quad \text{for } t \geq 0.001 \text{ sec}.$$

2-4. A 20.0 liter vessel initially contains air at 80.0% nitrogen and 20.0% oxygen. Nitrogen is added to the container at the constant rate of 0.1

liter/sec. Assume that continual mixing takes place in the vessel and the mixture of nitrogen and oxygen is withdrawn from the vessel at the rate at which nitrogen is added to the vessel. Develop a state model for the percentage of nitrogen in the vessel. What is the percentage of nitrogen in the vessel as a function of time?

2-5. A 100.0 liter beaker of water initially contains 1.0 kg of dissolved salt. Water is added at the constant rate of 5.0 liters/min with complete mixing occurring in the beaker; the salt solution is drained off at the same rate as water is added. Develop a state model for the mass of dissolved salt in the beaker. How much salt is in the beaker as a function of time? Assume that all salt in the beaker is in dissolved form.

2-6. A large chamber contains 200.0 meter^3 of a gas; initially 0.15% of the gas is carbon dioxide. A ventilator exchanges 20.0 meters^3/min of this gas with gas containing only 0.04% carbon dioxide. Give a state model for the percentage of carbon dioxide in the chamber. What is the percentage of carbon dioxide as a function of time?

2-7. A large container holds water which is maintained at a constant temperature of 20.0°C. A 0.5 kg mass of aluminum is added to the water; the initial temperature of the aluminum is 75.0°C. In 1.0 min the temperature of the aluminum is observed to decrease by 30.0°C. Develop a state model describing the temperature of the aluminum. Using the state model determine the temperature of the aluminum for $t \geqslant 0$.

2-8. A simple electrical circuit contains a 1.0 henry inductor in series with a 2.0 ohm resistor. What is a state model for the voltage across the resistor? How does the voltage across the resistor change as a function of time if initially there is 1.0 amp of current flowing in the circuit?

2-9. A simple electrical circuit contains a 1.0 ohm resistor in series with a 0.3 farad capacitor. Give a state model for the voltage across the capacitor. How does the voltage across the capacitor change as a function of time if initially the capacitor has a 2.0 coulomb charge on it?

2-10. Assume that the rate of increase of the population in a region is proportional to the population in that region. Assume the population doubles every 50.0 years. Give a state model for the population in the region. What is the population at a specified time if the initial population is 200.0 million individuals?

2-11. It is experimentally determined that a gram of radioactive radium decays by 0.44×10^{-3} gram in 1.0 year. Give a state model for the percentage of radium remaining. What is the percentage of radium remaining as a function of time?

2-12. A phase lead network contains a 10^{-4} farad capacitor and two 100.0 ohm resistors as shown below. The input is the ideal voltage source e_i; the output is the voltage drop e_0 across the indicated resistor.

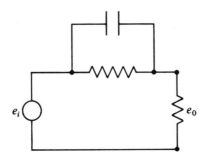

(a) Derive state equations for the circuit.
(b) What is the transfer function of the circuit?
(c) If the input voltage is $e_i(t) = 10.0$ volts for $t \geqslant 0$ what is the steady state output voltage?
(d) If the input voltage is $e_i(t) = 10.0 \cos 100t$ volts for $t \geqslant 0$ what is the steady state output voltage?

2-13. Consider the electrical circuit below with a 10^{-4} farad capacitor and two 100.0 ohm resistors.

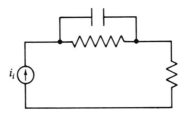

The input is the ideal current source i_i; the output is the current through the capacitor.

(a) Derive state equations.
(b) What is the transfer function of the circuit?
(c) If the input current is $i_i(t) = 2.0$ amps for $t \geqslant 0$ what is the steady state output current?
(d) If the input current is $i_i(t) = 2.0 \cos 100t$ amps for $t \geqslant 0$ what is the steady state output current?

Chapter III

Second Order Linear State Models

In this chapter physical processes are considered which can be described by second order linear state equations. A brief discussion of some mathematical aspects of such equations is given; then the systems approach is used to develop and analyze state models for several particular physical processes.

Some Systems Theory

In this section a state model of the form

$$\frac{dx_1}{dt} = a_{11}x_1 + a_{12}x_2 + b_1u,$$

$$\frac{dx_2}{dt} = a_{21}x_1 + a_{22}x_2 + b_2u,$$

$$y = c_1x_1 + c_2x_2$$

is examined; u is the input variable, y is the output variable and x_1, x_2 are the two state variables. The state model is defined by the constants $a_{11}, a_{12}, a_{21}, a_{22}, b_1, b_2, c_1, c_2$. An all-integrator block diagram of the abstract system represented by this state model is given in Figure 3-1.

These state equations are somewhat more complicated than the first order equations considered previously; hence it is somewhat more difficult to develop the corresponding theoretical results. But a few simple results can be presented.

Assuming the initial state is zero, taking the Laplace transforms of the

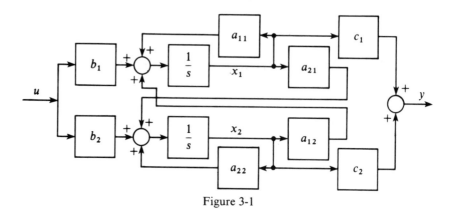

Figure 3-1

state equations leads to

$$sX_1 = a_{11}X_1 + a_{12}X_2 + b_1U,$$
$$sX_2 = a_{21}X_1 + a_{22}X_2 + b_2U,$$
$$Y = c_1X_1 + c_2X_2,$$

where the capital letters denote transforms of the corresponding variables. Thus

$$X_1 = \frac{\left[(s - a_{22})b_1 + a_{12}b_2\right]U}{(s - a_{11})(s - a_{22}) - a_{12}a_{21}},$$

$$X_2 = \frac{\left[a_{21}b_1 + (s - a_{11})b_2\right]U}{(s - a_{11})(s - a_{22}) - a_{12}a_{21}}$$

and

$$Y = \left\{\frac{c_1\left[(s - a_{22})b_1 + a_{12}b_2\right] + c_2\left[a_{21}b_1 + (s - a_{11})b_2\right]}{(s - a_{11})(s - a_{22}) - a_{12}a_{21}}\right\}U.$$

Consequently

$$G(s) = \frac{c_1\left[(s - a_{22})b_1 + a_{12}b_2\right] + c_2\left[a_{21}b_1 + (s - a_{11})b_2\right]}{(s - a_{11})(s - a_{22}) - a_{12}a_{21}}$$

is the transfer function; the transform of the zero state output response is the product of the transfer function and the transform of the input function. The denominator polynomial of the transfer function

$$d(s) = (s - a_{11})(s - a_{22}) - a_{12}a_{21}$$

is called the characteristic polynomial of the system. The zeros of this polynomial are called the characteristic zeros of the system; they are also the poles of the transfer function. It can be shown that if the real part of the characteristic zeros, which may be complex, are negative then the zero input response always tends to zero; the system is said to be stable. If this property does not hold then the system is said to be unstable.

Assuming that there are no common factors between the numerator polynomial and the denominator polynomial in the transfer function, it is clear that the state equations indicated, with two state variables, is a minimal state realization. One state variable would clearly not be sufficient to obtain a state realization while any state realization with more than two state variables would not be minimal.

It is quite difficult to develop explicit response expressions for the state equations indicated previously; in a particular case the determination of the response for a given initial state and given input function can be determined using Laplace transforms as usual.

To make the analysis somewhat simpler consider the special set of second order linear state equations, with constant parameters ξ, ω_0, b, defined by

$$\frac{dx_1}{dt} = x_2,$$

$$\frac{dx_2}{dt} = -2\xi\omega_0 x_2 - \omega_0^2 x_1 + bu,$$

$$y = x_1.$$

Such a special form occurs quite commonly and can be analyzed rather easily. The parameter ξ is usually referred to as the damping ratio of the system and ω_0 is referred to as the natural frequency of the system. The transfer function is given by

$$G(s) = \frac{b}{s^2 + 2\xi\omega_0 s + \omega_0^2}$$

and the characteristic polynomial is

$$d(s) = s^2 + 2\xi\omega_0 s + \omega_0^2.$$

Assuming $\xi > 0$ the system is always stable. If $0 < \xi < 1$ the characteristic zeros are complex; if $\xi > 1$ the characteristic zeros are real.

The zero input response is first examined; in this rather simple case it is not too difficult to determine the exact state responses using the method of Laplace transforms. Taking the transforms of the state equations obtain

$$sX_1 - x_1(0) = X_2,$$

$$sX_2 - x_2(0) = -2\xi\omega_0 X_2 - \omega_0^2 X_1,$$

so that

$$X_1 = \frac{(s + 2\xi\omega_0)x_1(0) + x_2(0)}{s^2 + 2\xi\omega_0 s + \omega_0^2},$$

$$X_2 = \frac{sx_2(0) - \omega_0^2 x_1(0)}{s^2 + 2\xi\omega_0 s + \omega_0^2}.$$

The state responses can be determined by taking the inverse transforms. The state response $x_1(t)$, if $0 < \xi < 1$, is given by

$$x_1(t) = x_1(0)e^{-\xi\omega_0 t}\cos\left(\omega_0\sqrt{1 - \xi^2}\,t\right)$$
$$+ \frac{x_2(0) + \xi\omega_0 x_1(0)}{\omega_0\sqrt{1 - \xi^2}}\,e^{-\xi\omega_0 t}\sin\left(\omega_0\sqrt{1 - \xi^2}\,t\right);$$

and if $\xi > 1$ the state response is given by

$$x_1(t) = \frac{-\omega_0\left(\xi - \sqrt{\xi^2 - 1}\,\right)x_1(0) - x_2(0)}{2\omega_0\sqrt{\xi^2 - 1}}\,e^{-\omega_0(\xi + \sqrt{\xi^2 - 1})t}$$
$$+ \frac{x_2(0) + \omega_0\left(\xi + \sqrt{\xi^2 - 1}\,\right)x_1(0)}{2\omega_0\sqrt{\xi^2 - 1}}\,e^{-\omega_0(\xi - \sqrt{\xi^2 - 1})t}.$$

Also $x_2(t) = (dx_1/dt)(t)$. Even in this simple case the zero input response is a rather complicated function of the initial state $x_1(0)$ and $x_2(0)$. In order to make this dependence more explicit it is convenient to plot, parametrically, for a fixed initial state one state function $x_2(t)$ versus the other state function $x_1(t)$. The x_1 versus x_2 plane is called the phase plane or the state plane; curves in this plane corresponding to solutions of the state equations are called trajectories. Some typical trajectories in the phase plane are indicated for two different values of the damping ratio in Figure 3-2. Note that if the damping ratio is less than one the trajectories are of the form of spirals since the characteristic zeros are complex; if the damping ratio exceeds one then the trajectories are of nodal type since the characteristic zeros are negative.

The arrows on the trajectories in Figure 3-2 indicate the change in the values of the state variables as time increases. Since each trajectory is parameterized by time, the time parameter could be indicated along the

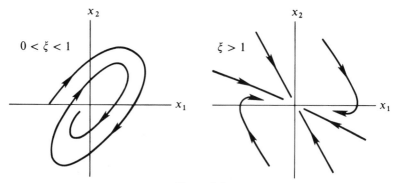

Figure 3-2

trajectory; this is usually omitted however. The phase plane is a rather simple graphical way of illustrating the qualitative dependence of the zero input response on the initial state. In this case the trajectories in the phase plane were determined from the time responses directly; this is usually very difficult. In many cases it is possible to determine, approximately, the form of the trajectories without use of the time responses. Use of the phase plane as a tool for understanding the zero input response is particularly valuable for systems which can be described by two state variables.

Now that discussion of the zero input response has been given, the zero state response is examined. The response for several standard input functions is determined.

First consider a constant input function $u(t) = \bar{u}$ for $t \geq 0$. Conditions for an equilibrium state are that $dx_1/dt = 0$ and $dx_2/dt = 0$ so that $x_1 = b\bar{u}/\omega_0^2, x_2 = 0$ define an equilibrium state. Thus if $x_1(0) = b\bar{u}/\omega_0^2$ and $x_2(0) = 0$ then $x_1(t) = b\bar{u}/\omega_0^2$ and $x_2(t) = 0$ for all $t \geq 0$; and $y(t) = b\bar{u}/\omega_0^2$ for all $t \geq 0$. If $0 < \xi < 1$ then the zero state response can be determined as

$$y(t) = \frac{b\bar{u}}{\omega_0^2} - \frac{b\bar{u}}{\omega_0^2} e^{-\xi\omega_0 t} \cos\left(\omega_0\sqrt{1-\xi^2}\,t\right)$$

$$- \frac{b\bar{u}\xi}{\omega_0^2\sqrt{1-\xi^2}} e^{-\xi\omega_0 t} \sin\left(\omega_0\sqrt{1-\xi^2}\,t\right),$$

so that $y(t) \to b\bar{u}/\omega_0^2$ as $t \to \infty$. If $\xi > 1$ then the zero state response can be determined as

$$y(t) = \frac{b\bar{u}}{\omega_0^2} + \frac{b\bar{u}}{2\omega_0^2\left(\xi^2 - 1 - \xi\sqrt{\xi^2-1}\right)} e^{-\omega_0(\xi - \sqrt{\xi^2-1})t}$$

$$+ \frac{b\bar{u}}{2\omega_0^2\left(\xi^2 - 1 + \xi\sqrt{\xi^2-1}\right)} e^{-\omega_0(\xi + \sqrt{\xi^2-1})t}$$

so that $y(t) \to b\bar{u}/\omega_0^2$ as $t \to \infty$. This clearly illustrates the fact that if $0 < \xi < 1$, so that the characteristic zeros are complex, then the responses include terms of the form

$$e^{-\xi\omega_0 t} \cos\left(\omega_0\sqrt{1-\xi^2}\,t\right) \quad \text{and} \quad e^{-\xi\omega_0 t}\sin\left(\omega_0\sqrt{1-\xi^2}\,t\right)$$

which represent a damped oscillation. If $\xi > 1$ so that the characteristic zeros are real, then the responses are purely exponential in form. If $0 < \xi < 1$ the response is said to be underdamped; if $\xi > 1$ the response is said to be overdamped; if $\xi = 1$ the response is said to be critically damped. The form of some typical responses to a constant input, showing the dependence on the damping ratio, are shown in Figure 3-3. The damping ratio $\xi = 0.707$ is generally considered to correspond to the "fastest" response; it represents a compromise between overshoot and rise time.

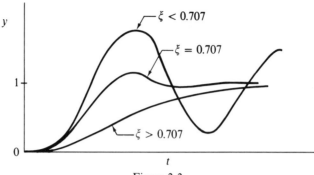

Figure 3-3

Now consider the impulse response function which is the inverse transform of the transfer function. If $0 < \xi < 1$, the impulse response function can be computed to be

$$y(t) = \frac{b}{\omega_0\sqrt{1 - \xi^2}} e^{-\xi\omega_0 t} \sin\left(\omega_0\sqrt{1 - \xi^2}\, t\right);$$

if $\xi > 1$ the impulse response function is

$$y(t) = \frac{b}{2\omega_0\sqrt{\xi^2 - 1}} \left[e^{-\omega_0(\xi - \sqrt{\xi^2 - 1})t} - e^{-\omega_0(\xi + \sqrt{\xi^2 - 1})t} \right].$$

Some typical impulse response functions, showing the dependence on the damping ratio, are shown in Figure 3-4.

Finally, consider the zero state response to a sinusoidal input function $u(t) = \bar{u}\cos\omega t$ for $t \geq 0$ for constant $\bar{u}$. Using the transfer function the transform of the zero state response is

$$Y = \frac{b\bar{u}s}{(s^2 + \omega^2)(s^2 + 2\xi\omega_0 s + \omega_0^2)}.$$

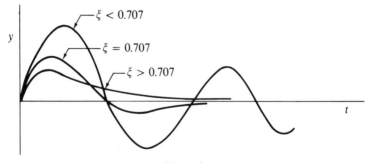

Figure 3-4

After some calculations it can be shown that

$$y(t) \to \frac{b\bar{u}}{\sqrt{\left(\omega^2 - \omega_0^2\right)^2 + 4\xi^2\omega_0^2\omega^2}} \cos\left[\omega t + \tan^{-1}\left(\frac{2\xi\omega_0\omega}{\omega^2 - \omega_0^2}\right)\right]$$

as $t \to \infty$. This last expression is referred to as the steady state response of the system to a sinusoidal input function. Note that the complex function $G(j\omega)$ can be written as

$$G(j\omega) = M(\omega)e^{j\phi(\omega)},$$

where

$$M(\omega) = \frac{b}{\sqrt{\left(\omega^2 - \omega_0^2\right)^2 + 4\xi^2\omega_0^2\omega^2}} \;,\quad \tan\phi(\omega) = \frac{2\xi\omega\omega_0}{\omega^2 - \omega_0^2}\;,$$

so that it is easily verified that the steady state response can be written as

$$y(t) = M(\omega)\bar{u}\cos(\omega t + \phi(\omega)).$$

Thus the steady state response to a sinusoidal input function can be determined directly from the frequency response function $G(j\omega)$. The amplitude $M(\omega)$ and phase $\phi(\omega)$, as a function of the damping ratio, are indicated in the Bode plots shown in Figure 3-5.

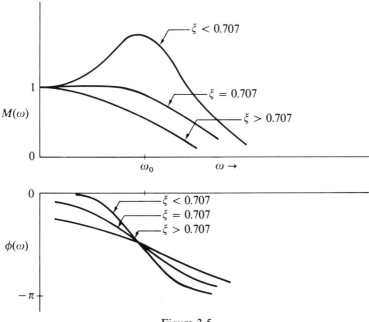

Figure 3-5

The zero input response and the zero state response have been characterized for a special class of second order linear systems. The general response is, of course, the appropriate sum of a zero state response and a zero input response.

A brief examination of second order linear systems has been given; a few theoretical systems results have been presented. Additional results are developed as required in the case study examples to follow.

Case Study 3-1: Ingestion and Metabolism of a Drug

In this example the ingestion and subsequent metabolism of a drug in a given individual are examined. A "two-compartment model" is used to characterize the ingestion, distribution and metabolism of the drug in the individual. In particular, the drug is ingested, e.g., orally as medication; the drug enters the gastrointestinal tract from where it is then distributed throughout the bloodstream of the individual to be metabolized and eliminated. This process is illustrated in Figure 3-1-1.

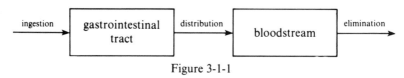

Figure 3-1-1

Let r denote the drug ingestion rate; let m_1 denote the mass of the drug in the first compartment, the gastrointestinal tract; let m_2 denote the mass of the drug in the second compartment, the bloodstream of the individual. The rate of change of the mass of drug in the gastrointestinal tract is equal to the rate at which the drug is ingested minus the rate at which the drug is distributed from the gastrointestinal tract to the bloodstream; this latter rate is assumed to be proportional to the mass of drug in the gastrointestinal tract. Hence

$$\frac{dm_1}{dt} = -K_1 m_1 + r,$$

where K_1 is a positive constant characterizing the gastrointestinal tract of the given individual. The rate of change of the mass of drug in the bloodstream is equal to the rate at which the drug is distributed from the gastrointestinal tract minus the rate at which the drug is metabolized and eliminated from the bloodstream; this latter rate is assumed to be proportional to the mass of drug in the bloodstream. Thus

$$\frac{dm_2}{dt} = K_1 m_1 - K_2 m_2.$$

The positive constant K_2 characterizes the metabolic and excretory processes of the individual. The above two differential equations constitute the two-compartment model for the ingestion, distribution and metabolism of a drug in an individual.

The ingestion rate r is naturally considered to be the input variable u; the output variable y is the mass m_2 of drug in the bloodstream since this is the variable which is indicative of the effect of the drug on the individual.

An obvious choice of state variables is given by the mass of drug in the gastrointestinal tract $x_1 = m_1$ and the mass of drug in the bloodstream $x_2 = m_2$.

Thus the state model is given by

$$\frac{dx_1}{dt} = -K_1 x_1 + u,$$

$$\frac{dx_2}{dt} = K_1 x_1 - K_2 x_2,$$

$$y = x_2.$$

The objective of this study is to determine how the input ingestion rate and the initial masses of drug in the body affect the subsequent amounts of drug in the bloodstream of the individual. The mathematical state model is used to determine the relationship between these variables.

The state model is a second order linear model. It should be noted that the first state equation does not include the second state variable; there is no feedback from the second state equation to the first state equation. The first state equation is said to be uncoupled from the second state equation. This observation is an indication of the fact that the mathematical model should not be excessively difficult to study analytically. This feature of the system structure can be indicated schematically using a block diagram, as shown in Figure 3-1-2. This block diagram is obtained directly from the state equations; the input, output, and state variables are indicated in the block diagram.

The mathematical model is now used to study the dynamic changes in the variables associated with the ingestion, distribution and metabolism of a drug in a given individual.

Suppose that there is an initial mass of drug in the gastrointestinal tract given by $x_1(0) = M_1$ and an initial mass of drug in the bloodstream given by $x_2(0) = M_2$. Assume an arbitrary ingestion rate $u(t)$, $t \geqslant 0$. Let X_1 denote the Laplace transform of x_1, X_2 denote the Laplace transform of x_2,

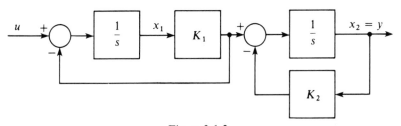

Figure 3-1-2

U denote the Laplace transform of u; from the state equations obtain

$$sX_1 - M_1 = -K_1X_1 + U,$$
$$sX_2 - M_2 = K_1X_1 - K_2X_2$$

so that

$$X_1 = \frac{M_1}{s + K_1} + \frac{U}{s + K_1},$$

$$X_2 = \frac{M_2}{s + K_2} + \frac{K_1M_1}{(s + K_1)(s + K_2)} + \frac{K_1U}{(s + K_1)(s + K_2)}.$$

Thus the transfer function from the input variable to the output variable is

$$G(s) = \frac{K_1}{(s + K_1)(s + K_2)}$$

and the characteristic polynomial, which is the denominator of the transfer function, is

$$d(s) = (s + K_1)(s + K_2).$$

The zeros of the characteristic polynomial, which are the poles of the transfer function, are both negative, namely $-K_1$ and $-K_2$. Thus, the conditions for stability are satisfied. Of course the damping ratio is necessarily greater than 1.0. The inverse transform of the transfer function is the impulse response function; it is determined, using a partial fraction expansion of $G(s)$, as

$$g(t) = \frac{K_1}{K_2 - K_1} \left[e^{-K_1 t} - e^{-K_2 t} \right],$$

where it is assumed that $K_1 \neq K_2$. Typically, the physiological constants satisfy the inequality

$$K_1 > K_2.$$

Several specific responses are now determined. First, consider the zero input response, where no drug is ingested, so that $u(t) = 0$, $t \geqslant 0$. Consequently, from the previous expressions for X_1 and X_2, it follows in this case that

$$x_1(t) = M_1 e^{-K_1 t},$$

$$x_2(t) = M_2 e^{-K_2 t}$$

$$+ \frac{K_1}{K_1 - K_2} \left[e^{-K_2 t} - e^{-K_1 t} \right]$$

so that $x_1(t) \to 0$, $x_2(t) \to 0$ as $t \to \infty$. These responses are consistent with the fact that the system is stable, as determined earlier. Since, in the usual case, $K_1 > K_2$ it follows that x_1 decays more rapidly than does x_2. Some typical time responses in this case are shown in Figure 3-1-3.

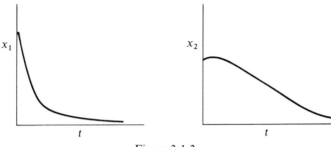

Figure 3-1-3

As indicated in the earlier discussion it is possible to examine the explicit dependence between variables x_1 and x_2 by exhibiting a typical trajectory in the phase plane. A qualitative picture of the trajectory corresponding to the time responses illustrated in Figure 3-1-3 is shown in Figure 3-1-4.

Now consider the special case where there is initially no drug in either the gastrointestinal tract or the bloodstream so that $M_1 = 0, M_2 = 0$. Also suppose that the drug ingestion rate is exponential as given by $u(t) = Re^{-\rho t}$, $t \geqslant 0$, where the ingestion rate is much greater than the distribution or metabolism rates for the drug, i.e., $\rho > K_1 > K_2$. Using the previously obtained expressions it follows that

$$X_1 = \frac{R}{(s + K_1)(s + \rho)},$$

$$X_2 = \frac{K_1 R}{(s + K_1)(s + K_2)(s + \rho)}.$$

After partial fraction expansions, the inverse transforms can be determined so that the zero state responses are

$$x_1(t) = \frac{R}{\rho - K_1}\left[e^{-K_1 t} - e^{-\rho t}\right],$$

$$X_2(t) = \frac{K_1 R}{(K_1 - K_2)(\rho - K_1)(\rho - K_2)}\left[(\rho - K_1)e^{-K_2 t}\right.$$

$$\left. - (\rho - K_2)e^{-K_1 t} + (K_1 - K_2)e^{-\rho t}\right].$$

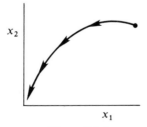

Figure 3-1-4

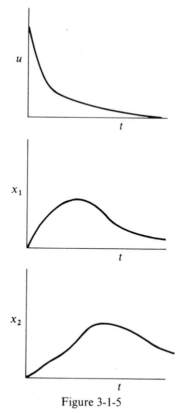

Figure 3-1-5

Typical time responses for the ingestion rate, the mass of drug in the gastrointestinal tract, the mass of drug in the bloodstream are shown in Figure 3-1-5.

Again, since $\rho > K_1 > K_2$, it takes a much longer period of time for the drug in the bloodstream to reach its maximum value and then decay than for the drug in the gastrointestinal tract. This is obviously consistent with the assumptions which were made in constructing the mathematical model.

Obviously, the model could be used to examine more complicated situations, e.g., the case of repeated ingestion of the drug. Analysis of such cases could be done using the analytical framework which has been developed.

A simple computer simulation program for this model, together with the simulation results, is given in Appendix E.

Case Study 3-2: Mixing of a Salt Solution

Two tanks contain a salt solution; assume L_1 and L_2 are the volumes of the two tanks and that the two tanks are interconnected as shown in Figure 3-2-1.

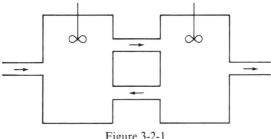

Figure 3-2-1

The solution in the first tank is pumped to the second tank at the constant flow rate Q_1; the solution in the second tank is pumped back to the first tank at the constant flow rate Q_2. Assume the net flow is from the first to the second tank. Assuming that each tank remains full the external flow rate into the first tank and out of the second tank is necessarily $Q_1 - Q_2$. It is assumed that there is a salt solution in each tank, at a uniform concentration due to complete mixing within each tank. The external stream added to the first tank contains a salt solution at a certain concentration c which is taken to be the input variable. The output is taken to be the concentration of salt withdrawn from the second tank. Our objective is to characterize the output concentration as a function of the input concentration and the initial amount of salt in each tank.

Let S_1 and S_2 be the mass of salt in each tank; then the total rate of change of mass of salt in the first tank is the rate at which salt is added to the first tank minus the rate at which salt is withdrawn from the first tank, i.e.,

$$\frac{dS_1}{dt} = -Q_1\left(\frac{S_1}{L_1}\right) + Q_2\left(\frac{S_2}{L_2}\right) + (Q_1 - Q_2)C.$$

Similarly, the total rate of change of mass of salt in the second tank is the rate at which salt is added to the second tank minus the rate at which salt is withdrawn from the second tank, i.e.,

$$\frac{dS_2}{dt} = Q_1\left(\frac{S_1}{L_1}\right) - Q_2\left(\frac{S_2}{L_2}\right) - (Q_1 - Q_2)\left(\frac{S_2}{L_2}\right).$$

The above two equations describing the relation between physical variables in the process have been determined from physical considerations, namely that mass is conserved. It is assumed that $Q_1 > 0, Q_2 > 0$ and that $Q_1 - Q_2 > 0$. Physically, it is required that $C \geqslant 0$ and that $S_1 \geqslant 0$ and $S_2 \geqslant 0$, always.

Clearly the masses of salt in each tank serve as state variables so define $x_1 = S_1$ and $x_2 = S_2$; the input variable u is the concentration C of salt added to the first tank; the output variable y is the concentration of salt in the second tank, S_2/L_2. Thus the state equations for the physical process

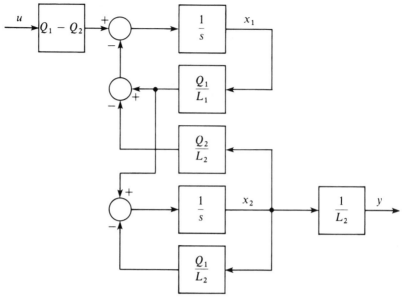

Figure 3-2-2

are

$$\frac{dx_1}{dt} = -\left(\frac{Q_1}{L_1}\right)x_1 + \left(\frac{Q_2}{L_2}\right)x_2 + (Q_1 - Q_2)u,$$

$$\frac{dx_2}{dt} = \left(\frac{Q_1}{L_1}\right)x_1 - \left(\frac{Q_1}{L_2}\right)x_2,$$

$$y = \left(\frac{1}{L_2}\right)x_2.$$

This is clearly a linear second order system. An all-integrator block diagram is shown in Figure 3-2-2.

The responses of this system are now examined in some detail.

First, consider the zero input response where $u(t) = 0$ for $t \geqslant 0$, i.e., the external flow to the first tank contains no salt. Thus, taking the transform of the state equations

$$sX_1 - M_1 = -\left(\frac{Q_1}{L_1}\right)X_1 + \left(\frac{Q_2}{L_2}\right)X_2,$$

$$sX_2 - M_2 = \left(\frac{Q_1}{L_1}\right)X_1 - \left(\frac{Q_1}{L_2}\right)X_2,$$

where M_1 and M_2 are the initial masses of salt in each tank; then

$$X_1 = \frac{(s + (Q_1/L_2))M_1 + (Q_2/L_2)M_2}{s^2 + ((Q_1/L_1) + (Q_1/L_2))s + (Q_1(Q_1 - Q_2)/L_1L_2)},$$

$$X_2 = \frac{(Q_1/L_1)M_1 + (s + (Q_1/L_1))M_2}{s^2 + ((Q_1/L_1) + (Q_1/L_2))s + (Q_1(Q_1 - Q_2)/L_1L_2)}.$$

The state responses can be obtained by determining the inverse transforms. The characteristic polynomial is

$$d(s) = s^2 + \left(\frac{Q_1}{L_1} + \frac{Q_1}{L_2} \right)s + \frac{Q_1(Q_1 - Q_2)}{L_1 L_2} \; ;$$

in this case it can be factored as $d(s) = (s - \lambda_1)(s - \lambda_2)$ where

$$\lambda_1 = \frac{1}{2} \left\{ -\left(\frac{Q_1}{L_1} + \frac{Q_1}{L_2} \right) + \sqrt{ \left(\frac{Q_1}{L_1} - \frac{Q_1}{L_2} \right)^2 + \frac{4 Q_1 Q_2}{L_1 L_2} } \right\},$$

$$\lambda_2 = \frac{1}{2} \left\{ -\left(\frac{Q_1}{L_1} + \frac{Q_1}{L_2} \right) - \sqrt{ \left(\frac{Q_1}{L_1} - \frac{Q_1}{L_2} \right)^2 + \frac{4 Q_1 Q_2}{L_1 L_2} } \right\}$$

where, since $Q_1 > Q_2$, it follows that λ_1 and λ_2 are both negative and unequal. The system is thus stable with a damping ratio greater than one. The state responses thus are

$$x_1(t) = \left[\left(\frac{Q_1}{L_2} \right)M_1 + \left(\frac{Q_2}{L_2} \right)M_2 \right]\left[\frac{e^{\lambda_2 t} - e^{\lambda_1 t}}{\lambda_2 - \lambda_1} \right]$$
$$+ M_1 \left[\frac{\lambda_2 e^{\lambda_2 t} - \lambda_1 e^{\lambda_1 t}}{\lambda_2 - \lambda_1} \right],$$

$$x_2(t) = \left[\left(\frac{Q_1}{L_1} \right)M_1 + \left(\frac{Q_1}{L_1} \right)M_2 \right]\left[\frac{e^{\lambda_2 t} - e^{\lambda_1 t}}{\lambda_2 - \lambda_1} \right]$$
$$+ M_2 \left[\frac{\lambda_2 e^{\lambda_2 t} - \lambda_1 e^{\lambda_1 t}}{\lambda_2 - \lambda_1} \right].$$

Since $\lambda_1 < 0$ and $\lambda_2 < 0$ then $x_1(t) \to 0$ and $x_2(t) \to 0$ as $t \to \infty$. Hence $y(t) \to 0$ as $t \to \infty$. In physical terms, if external flow into the first tank contains no salt then the system is eventually flushed of all salt within the two tanks.

Some trajectories in the phase-plane can be easily sketched. Note that the slope of a trajectory is given by

$$\frac{dx_2}{dx_1} = \frac{dx_2/dt}{dx_1/dt} = \frac{(Q_1/L_1)x_1 - (Q_1/L_2)x_2}{-(Q_1/L_1)x_1 + (Q_2/L_2)x_2} \; ;$$

this relationship is particularly valuable as a means of obtaining sufficient graphical information to sketch the trajectories in the phase plane. For if $-(Q_1/L_1)x_1 + (Q_2/L_2)x_2 = 0$ then $dx_2/dx_1 = \infty$ and necessarily $dx_2/dt = ((Q_2 - Q_1)/L_2)x_2$. Now if $(Q_1/L_1)x_1 - (Q_1/L_2)x_2 = 0$ then $dx_2/dx_1 = 0$ and necessarily $dx_1/dt = ((Q_2 - Q_1)/L_1)x_1$. Thus each trajectory crosses the straight line $-(Q_1/L_1)x_1 + (Q_2/L_2)x_2 = 0$ vertically with dx_2/dt opposite in sign to x_2. Each trajectory crosses the straight line $(Q_1/L_1)x_1 - (Q_1/L_2)x_2 = 0$ horizontally with dx_1/dt opposite in sign to x_1. Note also that the state equations indicate that if $-(Q_1/L_1)x_1 +$

Figure 3-2-3

$(Q_2/L_2)x_2 > 0$ then $dx_1/dt > 0$ and vice-versa; and if $(Q_1/L_1)x_1 - (Q_1/L_2)x_2 > 0$ then $dx_2/dt > 0$ and vice-versa. All of this information can be used to sketch, approximately, some trajectories in the phase-plane. Of course, as shown previously, all trajectories tend to the equilibrium solution at $x_1 = 0$ and $x_2 = 0$. Some typical trajectories in the phase plane are shown in Figure 3-2-3. Of course the state variables are of interest, physically, only in the first quadrant where $x_1 \geqslant 0, x_2 \geqslant 0$.

Now consider the zero state response of the system, i.e., where there is initially no salt in either tank. Based on the state equations the transfer function can easily be determined to be

$$G(s) = \frac{Q_1(Q_1 - Q_2)/L_1L_2}{d(s)},$$

where the characteristic polynomial is as before.

Suppose the concentration of salt in the external flow to the first tank is constant, say $u(t) = \overline{C}$ for $t \geqslant 0$. From the state equations it follows that $dx_1/dt = 0$ and $dx_2/dt = 0$ when $x_1 = L_1\overline{C}$ and $x_2 = L_2\overline{C}$; thus $x_1 = L_1\overline{C}$ and $x_2 = L_2\overline{C}$ is an equilibrium state; thus $y(t) = \overline{C}$ for $t \geqslant 0$ is the corresponding constant output. Now the transform of the zero state output response is

$$Y = \left(\frac{Q_1(Q_1 - Q_2)/L_1L_2}{sd(s)} \right)\overline{C};$$

After a partial fraction expansion the inverse transform can be obtained to yield the zero state response

$$y(t) = \overline{C} + \frac{Q_1(Q_1 - Q_2)}{L_1L_2}\overline{C}\left\{ \frac{e^{\lambda_1 t}}{\lambda_1(\lambda_1 - \lambda_2)} + \frac{e^{\lambda_2 t}}{\lambda_2(\lambda_2 - \lambda_1)} \right\}.$$

Again, it is seen that

$$y(t) \to \overline{C} \quad \text{as } t \to \infty.$$

That is, the concentration of salt in the second tank tends to the concentration of salt in the flow added to the first tank. Clearly, this occurs regardless of the initial mass of salt in the two tanks.

Case Study 3-3: DC Motor

Our objective in this case study is to present the operational considerations for a general DC motor, to translate those assumptions into mathematical state equations, and finally to analyze those state equations as a means of obtaining insight into the response characteristics of DC motors. No detailed examination of DC motor construction is attempted; rather a general overview of the relevant assumptions about the operation of DC motors is given. The detailed analysis of DC motors is carried out in two special but important cases: a constant armature current and a constant field current.

A DC motor is essentially a device for converting electrical energy into mechanical, or rotational, energy. The basic ingredients of the motor are an armature circuit and a field circuit; these circuits are coupled in such a way that a torque is produced on the motor shaft. A schematic picture of this process is indicated in Figure 3-3-1.

The voltage applied to the field and armature circuits are v_f and v_a respectively; i_f and i_a are the current in the field and armature circuits respectively. The torque generated by the motor is T_e and ω is the motor speed. In the field circuit

$$- v_f + R_f i_f + L_f \frac{di_f}{dt} = 0,$$

where R_f and L_f are the field resistance and inductance. In the armature circuit there is resistance R_a and inductance L_a; there is also a voltage drop in the armature circuit which is proportional to the product of the motor speed and the field current; thus

$$- v_a + R_a i_a + L_a \frac{di_a}{dt} + B\omega i_f = 0,$$

where B is an electromechanical constant for the motor. The field current is clearly independent of the armature current but the armature current does depend on the field current and also the motor speed as indicated. Finally, the torque T_e generated by the motor is assumed to be proportional to the product of the field current and the armature current according to the

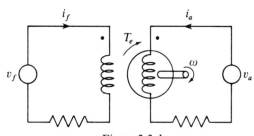

Figure 3-3-1

equation

$$T_e = B i_a i_f.$$

In terms of deriving a realistic model for the motor it is necessary to consider the rotational motion of the motor shaft. Here it is assumed that the inertia of the motor shaft, as well as any external load, is J; and that there is a viscous damping torque which is proportional to the motor speed, where c is the damping coefficient. Thus

$$J \frac{d\omega}{dt} = -c\omega + T_e.$$

The above four mathematical equations constitute the basis for the model of the DC motor. The external input variables are the applied field and/or armature voltages; the variables of interest are the field and/or armature currents and the motor speed. A careful analysis of the above equations is quite difficult since several of the equations involve products of the variables and hence are nonlinear equations. In what follows two special but important cases are examined. The first case corresponds to the situation where the armature current is a constant; the second corresponds to a constant field current. The physical mechanisms by which these two special cases are obtained is not considered.

Constant Armature Current. In this section it is assumed that the armature current is maintained at a constant value, namely

$$i_a = I_a.$$

Such operation is referred to as a field controlled DC motor. From the previous equations it follows that

$$-v_f + R_f i_f + L_f \frac{d i_f}{dt} = 0$$

and

$$J \frac{d\omega}{dt} = -c\omega + B I_a i_f$$

are the descriptive equations for the motor. It is an easy matter to choose state variables as the field current $x_1 = i_f$ and the motor speed $x_2 = \omega$. The input variable u is the applied field voltage v_f; the output variable y is the motor speed ω.

The state equations can be written as

$$\frac{dx_1}{dt} = -\frac{R_f}{L_f} x_1 + \frac{u}{L_f},$$

$$\frac{dx_2}{dt} = -\frac{c}{J} x_2 + \left(\frac{B I_a}{J} \right) x_1,$$

which is a second order linear set of state equations for the field controlled DC motor. The block diagram corresponding to the above state equations is shown in Figure 3-3-2.

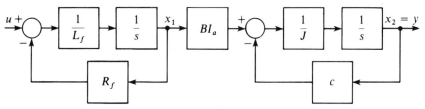

Figure 3-3-2

In the state equations the first state equation is said to be uncoupled; in terms of block diagram this corresponds to the fact that there is no feedback from the second state variable to the first state variable. In physical terms, the motion of the shaft does not affect the current in the field. In such cases where one or more state equations are uncoupled the analysis is made considerably easier.

First, consider the zero input response with no applied field voltage, i.e., $u(t) = 0$, $t \geqslant 0$. Taking the transforms of the state equations

$$sX_1 - I_f = -\frac{R_f}{L_f}X_1,$$

$$sX_2 - \Omega = -\frac{c}{J}X_2 + \left(\frac{BI_a}{J}\right)X_1,$$

where I_f and Ω are the initial field current and motor speed. Thus

$$X_1 = \frac{I_f}{s + (R_f/L_f)}$$

$$X_2 = \frac{(BI_a/J)I_f + (s + (R_f/L_f))\Omega}{(s + (c/J))(s + (R_f/L_f))}.$$

The characteristic polynomial is

$$d(s) = \left(s + \frac{c}{J}\right)\left(s + \frac{R_f}{L_f}\right).$$

In this case the characteristic polynomial and hence the characteristic zeros do not depend on the electromechanical constant B. Assuming that all constants are positive the DC motor is necessarily stable; the damping ratio is easily computed to be

$$\xi = \frac{1}{2}\left[\sqrt{\frac{c}{J}\frac{L_f}{R_f}} + \sqrt{\frac{J}{c}\frac{R_f}{L_f}}\right].$$

Simple algebra shows that $\xi > 1$ so that the zero input response tends exponentially to zero as expected. The time responses for the state variables could be determined by taking the inverse transform of the above expressions, but the nature of the responses is best illustrated by examination of corresponding trajectories in the phase plane. The slope of each trajectory is given by the expression

$$\frac{dx_2}{dx_1} = \frac{dx_2/dt}{dx_1/dt} = \frac{-(c/J)x_2 + (BI_a/J)x_1}{-(R_f/L_f)x_1}.$$

This expression is very useful as a means of determining the qualitative form of the trajectories without first having to determine the time responses. The following observations are easily made. If $-cx_2 + BI_a x_1 = 0$ then $dx_2/dx_1 = 0$ so that a trajectory through such x_1, x_2 in the phase plane has zero slope, i.e., it is horizontal; note further that dx_1/dt and x_1 are of opposite sign. Now if $x_1 = 0$ then $dx_2/dx_1 = \infty$ so that a trajectory through such x_1, x_2 in the phase plane has infinite slope, i.e., it is vertical; and dx_2/dt and x_2 are of opposite sign. Now if $(cL_f - JR_f)x_2 = (BI_a L_f)x_1$ then the slope of a trajectory through such x_1, x_2 is given by

$$\frac{dx_2}{dx_1} = \frac{BI_a L_f}{cL_f - JR_f},$$

i.e., the slope of the trajectory is the same as the slope of the line; the line is necessarily a trajectory itself; along this line dx_1/dt and x_1 are of opposite sign. Now it is possible to use all of this information to sketch some typical trajectories in the phase plane. This is done in Figure 3-3-3, assuming that $cL_f - JR_f > 0$.

It is clear that all trajectories tend to the equilibrium state $x_1 = 0, x_2 = 0$. A given initial state defines a trajectory in the phase plane; this trajectory could be used to make qualitative inferences about the corresponding zero input time responses.

The zero state response is now examined; hence assume that $x_1(0) = 0$, $x_2(0) = 0$. For simplicity only the motor speed is analyzed. Using the state equations, the transforms X_1 and X_2 and U are related by

$$sX_1 = -\frac{R_f}{L_f}X_1 + \frac{U}{L_f},$$

$$sX_2 = -\frac{c}{J}X_2 + \left(\frac{BI_a}{J}\right)X_1.$$

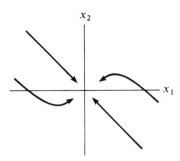

Figure 3-3-3

The transfer function from the field voltage to the motor speed is thus

$$G(s) = \frac{BI_a/JL_f}{(s + (c/J))(s + (R_f/L_f))} .$$

The impulse response function is the inverse transform of the transfer function. After a partial fraction expansion the impulse response function is determined to be

$$g(t) = \frac{BI_a}{JR_f - cL_f} \left[\exp\left(-\frac{c}{J}t\right) - \exp\left(-\frac{R_f}{L_f}t\right) \right].$$

Thus the zero state motor speed response can be expressed in terms of the applied field voltage by the equation

$$x_2(t) = \frac{BI_a}{JR_f - cL_f} \int_0^t \left[\exp\left(-\frac{c}{J}(t - \tau)\right) - \exp\left(-\frac{R_f}{L_f}(t - \tau)\right) \right] u(\tau)\, d\tau.$$

In the particular case of a constant field voltage $u(t) = E_f$ for $t \geqslant 0$, the zero state response can be calculated to be

$$x_2(t) = \frac{BI_a E_f}{cR_f} + \frac{BI_a E_f}{((R_f/L_f) - (c/J))} \left[\frac{1}{JR_f} \exp\left(-\frac{R_f}{L_f}t\right) \right.$$

$$\left. - \frac{1}{cL_f} \exp\left(-\frac{c}{J}t\right) \right].$$

Thus

$$x_2(t) \to \frac{BI_a E_f}{cR_f} \quad \text{as } t \to \infty$$

which corresponds to the equilibrium state

$$x_1 = \frac{E_f}{R_f}, \quad x_2 = \frac{BI_a E_f}{cR_f} .$$

Since the DC motor is stable it follows that for any initial state

$$x_2(t) \to \frac{BI_a E_f}{cR_f} \quad \text{as } t \to \infty.$$

Thus, if a constant voltage is applied to the field then, no matter what the initial current in the field and the initial motor speed, the angular velocity of the shaft tends exponentially to the constant value given above. Note that the steady state motor speed given above does not depend on the inertial load J; however, one of the characteristic zeros, $-(c/J)$, does depend on the load. In particular for large loads, i.e., large values of J, the motor may require a relatively long time to approach its steady state speed.

Constant Field Current. In this section it is assumed that the field current is maintained at a constant value

$$i_f = I_f.$$

Such operation is referred to as an armature controlled DC motor. From the original description it follows that the armature current i_a and the motor speed ω satisfy the equations

$$-v_a + R_a i_a + L_a \frac{di_a}{dt} + (BI_f)\omega = 0,$$

$$J \frac{d\omega}{dt} = -c\omega + (BI_f)i_a.$$

The armature current and motor speed can be chosen as state variables so that $x_1 = i_a$ and $x_2 = \omega$; the input variable u is the applied armature voltage v_a; the output variable y is the motor speed ω.

Thus the state equations are

$$\frac{dx_1}{dt} = -\frac{R_a}{L_a}x_1 - \left(\frac{BI_f}{L_a}\right)x_2 + \frac{u}{L_a},$$

$$\frac{dx_2}{dt} = -\frac{c}{J}x_2 + \left(\frac{BI_f}{J}\right)x_1.$$

This is again a second order linear state model for the armature controlled DC motor. The block diagram corresponding to these state equations is shown in Figure 3-3-4.

In these state equations, the derivative of each state variable depends explicitly on both state variables; hence the state equations are coupled. This coupling is exhibited in the block diagram as feedback. Since the state equations are coupled they must be analyzed simultaneously.

First, consider the zero input response with no applied armature voltage, i.e., $u(t) = 0$, $t \geqslant 0$. Taking the transforms of the state equations

$$sX_1 - I_a = -\frac{R_a}{L_a}X_1 - \left(\frac{BI_f}{L_a}\right)X_2,$$

$$sX_2 - \Omega = -\frac{c}{J}X_2 + \left(\frac{BI_f}{J}\right)X_1,$$

where I_a and Ω are the initial armature current and motor speed. Solving for X_1 and X_2 obtain

$$X_1 = \frac{(s + (c/J))I_a - (BI_f/L_a)\Omega}{d(s)},$$

$$X_2 = \frac{(BI_f/J)I_a + (s + (R_a/L_a))\Omega}{d(s)},$$

where the characteristic polynomial is

$$d(s) = \left(s + \frac{c}{J}\right)\left(s + \frac{R_a}{L_a}\right) + \frac{B^2 I_f^2}{JL_a}.$$

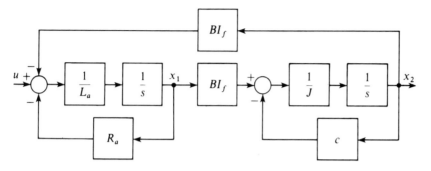

Figure 3-3-4

In this case the characteristic polynomial does depend on the electro-mechanical constant B. The characteristic polynomial is not obtained in factored form but it could be factored using the quadratic formula. Depending on the particular values of the parameters the characteristic zeros could be real or complex. In either case the DC motor is necessarily stable. The two cases are best distinguished by computing the damping ratio which is

$$\xi = 0.5 \sqrt{\frac{JL_a}{cR_a + B^2 I_f^2}} \left[\frac{c}{J} + \frac{R_a}{L_a} \right].$$

It is clear that, depending on the parameter values, the damping ratio can be either smaller than or greater than unity. The character of the zero input responses differ in the two cases; if $0 < \xi < 1$ the zero input responses are damped oscillations, if $\xi > 1$ the zero input responses are exponential. These two cases could be considered in some detail. In each case the time responses could be determined from the previous transforms. Trajectories in the phase plane could be determined using the techniques discussed elsewhere.

The zero state response is now examined; suppose that $x_1(0) = 0$, $x_2(0) = 0$. Using the state equations, the transforms X_1 and X_2 and U are related by

$$sX_1 = -\frac{R_a}{L_a} X_1 - \left(\frac{BI_f}{L_a} \right) X_2 + \frac{U}{L_a},$$

$$sX_2 = -\frac{c}{J} X_2 + \left(\frac{BI_f}{J} \right) X_1.$$

The primary interest here is in the motor speed as it depends on the input armature voltage; the above simultaneous equations can be solved to obtain the transfer function from armature voltage to motor speed

$$G(s) = \frac{BI_f/JL_a}{d(s)},$$

where $d(s)$ is the characteristic polynomial given previously in this section. The impulse response is the inverse transform of $G(s)$; the particular form of the impulse response function depends on whether $0 < \xi < 1$ or $\xi > 1$. In the particular case of a constant armature voltage, $u(t) = E_a$ for $t \geqslant 0$, the zero state response could be determined by taking the inverse transform of

$$X_2 = \frac{BI_f / JL_a}{d(s)} \frac{E_a}{s}.$$

It is an easy matter to observe that

$$X_2 = \frac{BI_f E_a}{cR_a + B^2 I_f^2} \frac{1}{s} + \frac{p(s)}{d(s)},$$

where $p(s)$ is a polynomial in s. Since the zeros of $d(s)$ all have negative real parts it follows that the zero state motor speed tends asymptotically to a constant value. In fact for any initial state it follows that the motor speed satisfies

$$x_2(t) \to \frac{BI_f E_a}{cR_a + B^2 I_f^2} \quad \text{as } t \to \infty.$$

It can be seen from the state equations that this steady state response corresponds to the equilibrium state

$$x_1 = \frac{cE_a}{cR_a + B^2 I_f^2}, \quad x_2 = \frac{BI_f E_a}{cR_a + B^2 I_f^2}.$$

Thus, if a constant voltage is applied to the armature then no matter what the initial current in the armature and the initial motor speed the angular velocity of the shaft tends asymptotically to the constant value given above. The details of the transient response depend on the particular value of the damping ratio, whether $0 < \xi < 1$ or $\xi > 1$. The steady state motor speed does not depend on the inertial load J; however the terms in the characteristic polynomial do depend on J in a way that for large loads, i.e., large values of J, the motor may require a relatively long time to approach its steady state speed.

Case Study 3-4: A Bridged-T-Filter

An important example of an electrical network used as a filter in many amplitude modulated communication devices is a bridged -T- filter of the form shown in Figure 3-4-1.

The circuit consists of two resistors with resistances R and nR, where $n > 1$, and two capacitors with capacitances C and mC as shown in Figure 3-4-1. Such circuits are used to attenuate certain undesired harmonic signals in the communication device. The input to the circuit is the voltage

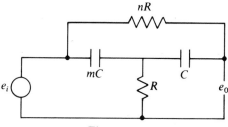

Figure 3-4-1

source e_i, assumed to have zero impedance; e_0 is the output voltage of the circuit, assumed to be determined with infinite impedance. Our objective in this example is to determine the relation between the input voltage and the output voltage.

The equations describing the circuit are first obtained. Suppose that I_1 and I_2 are loop currents, assumed clockwise; also suppose that Q_1 is the charge on the capacitor with capacitance C; and Q_2 is the charge on the capacitor of capacitance mC. Then the voltage drop in each loop is necessarily zero, i.e.,

$$-e_i + \frac{Q_2}{mC} + RI_1 = 0,$$

$$-\frac{Q_2}{mC} + nRI_2 + \frac{Q_1}{C} = 0.$$

The charge-current relation through each capacitor is

$$\frac{dQ_1}{dt} = I_2,$$

$$\frac{dQ_2}{dt} = I_1 - I_2.$$

These are the physical equations, based on Kirchoff's laws, which characterize the circuit.

There are many possible choices of state variables in this circuit; the most natural is to choose the charges on each capacitor as the state variables, i.e., $x_1 = Q_1$ and $x_2 = Q_2$. The input variable u is the voltage e_i; the output variable y is the output voltage $e_0 = (Q_1/C) - (Q_2/mC) + e_i$. Thus the state model is given by

$$\frac{dx_1}{dt} = -\frac{x_1}{nRC} + \frac{x_2}{mnRC},$$

$$\frac{dx_2}{dt} = \frac{x_1}{nRC} - \left(\frac{1}{mnRC} + \frac{1}{mRC}\right)x_2 + \frac{u}{R},$$

$$y = \frac{x_1}{C} - \frac{x_2}{mC} + u.$$

These are linear state equations involving two state variables.

The transfer function is first determined. Assuming the initial state is zero the transform of the two state equations leads to

$$sX_1 = -\frac{X_1}{nRC} + \frac{X_2}{mnRC} ,$$

$$sX_2 = \frac{X_1}{nRC} - \left(\frac{1}{mnRC} + \frac{1}{mRC}\right)X_2 + \frac{U}{R} .$$

After careful calculation the transforms X_1 and X_2 can be determined in terms of the transform of the input U as

$$X_1 = \frac{CU}{mnR^2C^2s^2 + (1 + m + n)RCs + 1} ,$$

$$X_2 = \frac{m(1 + nRCs)CU}{mnR^2C^2s^2 + (1 + m + n)RCs + 1} .$$

Using the expression for the output in terms of the state variables the transform Y can be expressed in terms of the transform U to obtain the transfer function from u to y as

$$G(s) = \frac{mnR^2C^2s^2 + (1 + m)RCs + 1}{mnR^2C^2s^2 + (1 + m + n)RCs + 1} .$$

The characteristic polynomial is

$$d(s) = mnR^2C^2s^2 + (1 + m + n)RCs + 1$$

so that the characteristic zeros have negative real parts and hence the system is stable. Note also that since the characteristic polynomial is of second degree at least two state variables are required so that the given state equations, with two state variables, is minimal.

Some of the response properties of the system are now determined. Since the system is stable it follows that the zero input response, i.e., the response to zero input voltage, always tends to zero for any initial state. To examine the zero input response in more detail the trajectories in the phase plane can be examined. With $u(t) = 0$ for $t \geqslant 0$, the expression for the slope of the trajectories

$$\frac{dx_2}{dx_1} = \frac{dx_2/dt}{dx_1/dt} = \frac{mx_1 - (n + 1)x_2}{-mx_1 + x_2}$$

can be obtained from the state equations. This equation can be used as a basis for determining sufficient information about the trajectories so that they can be roughly sketched in the phase plane. For example, the following information is easily determined: if $x_1 = 0$ then $dx_2/dx_1 = (n - 1) > 0$; if $x_2 = 0$ then $dx_2/dx_1 = -1$; let k denote either the positive or negative solution of the quadratic $k^2 + (1 + n - m)k - m = 0$; then the lines defined by $x_2 = kx_1$ define particular trajectories, called asymptotes. The general expression for the slope of the trajectories could be used to

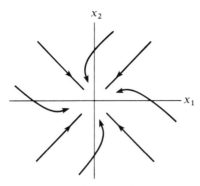

Figure 3-4-2

obtain additional information as well. Also recall that, from the state equations, dx_1/dt has the same sign as $x_2 - mx_1$ and dx_2/dt has the same sign as $mx_1 - (1 + n)x_2$; this information is useful for determining regions in the phase plane where x_1 and x_2 increase with time, regions where x_1 increases and x_2 decreases with time, etc. These simple observations can now be used to sketch, qualitatively, some typical trajectories in the phase plane, as shown in Figure 3-4-2.

The typical trajectories illustrated show very nicely the qualitative dependence of the state responses on the initial states; the output response can easily be inferred. Of course the fact that all trajectories tend to the zero equilibrium state is consistent with the fact that the system is stable, as seen from examination of the characteristic polynomial.

The most interesting class of voltage input functions are those which are sinusoidal, since such functions are of supreme importance in most communication devices. Hence the responses of the circuit, in the case $u(t) = \bar{u}\cos\omega t$ for $t \geqslant 0$, are now examined. The transform of the zero state response is given by

$$Y = \frac{mnR^2C^2s^2 + (1 + m)RCs + 1}{mnR^2C^2s^2 + (1 + m + n)RCs + 1} \frac{\bar{u}s}{s^2 + \omega^2},$$

and it is known that the inverse transform satisfies

$$y(t) \rightarrow M(\omega)\bar{u}\cos(\omega t + \phi(\omega)) \quad \text{as } t \rightarrow \infty,$$

where

$$M(\omega)e^{j\phi(\omega)} = G(j\omega) = \frac{(1 - mnR^2C^2\omega^2) + j(1 + m)RC\omega}{(1 - mnR^2C^2\omega^2) + j(1 + m + n)RC\omega}.$$

It is convenient to introduce the parameter

$$\omega_0^2 = \frac{1}{mnR^2C^2}.$$

After some complex algebra obtain

$$M(\omega) = \sqrt{\frac{mn(\omega_0^2 - \omega^2)^2 + (1+m)^2\omega_0^2\omega^2}{mn(\omega_0^2 - \omega^2)^2 + (1+m+n)^2\omega_0^2\omega^2}}$$

and

$$\tan\phi(\omega) = \frac{-n\sqrt{nm}\,(\omega_0^2 - \omega^2)\omega\omega_0}{mn(\omega_0^2 - \omega^2)^2 + \omega^2\omega_0^2(1+m)(1+m+n)}.$$

These frequency response functions have the following properties: if $\omega = 0$ then $M(\omega) = 1$ and $\phi(\omega) = 0$; if $0 < \omega < \omega_0$ then $(1 + m/1 + m + n) < M(\omega) < 1$ and $\phi(\omega) < 0$; if $\omega = \omega_0$ then $M(\omega) = (1 + m)/(1 + m + n)$ and $\phi(\omega) = 0$; if $\omega_0 < \omega$ then $(1 + m)/(1 + m + n) < M(\omega) < 1$ and $\phi(\omega) > 0$; as $\omega \to \infty$ then $M(\omega) \to 1$ and $\phi(\omega) \to 0$. Thus the qualitative form of the Bode plot is as shown in Figure 3-4-3.

Thus for a sinusoidal input of the form $u(t) = \bar{u}\cos\omega t$ for $t \geqslant 0$, the response of the circuit for any initial state satisfies

$$y(t) \to M(\omega)\bar{u}\cos(\omega t + \phi(\omega)) \quad \text{as } t \to \infty,$$

where $M(\omega)$ and $\phi(\omega)$ are as above. The important fact to note is that if $\omega = \omega_0$ then the steady state output response is $y(t) = (1 + m)/(1 + m + n)$ $\bar{u}\cos\omega_0 t$ for $t \geqslant 0$. Hence if n is much greater than unity, as is usual, then the output is necessarily attenuated. For input frequencies ω near ω_0 there is lesser attenuation while for $\omega = 0$ and $\omega \to \infty$ the output voltage is the same as the input voltage. This particular circuit is sometimes referred to as a notch-type filter where $\omega_0 = (\sqrt{mn}\,RC)^{-1}$ is the notch frequency at which the voltage transmission through the circuit is a minimum.

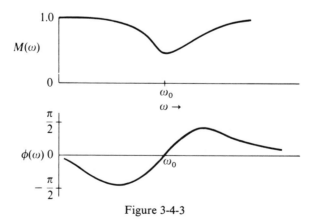

Figure 3-4-3

Exercises

3-1. Consider the model for ingestion and metabolism of a drug in an individual given in Case Study 3-1. Assume the parameter values

$$K_1 = 0.03 \; 1/\text{min}, \quad K_2 = 0.01 \; 1/\text{min}.$$

(a) Assume no drug is ingested so the ingestion rate is $r(t) = 0$ for $t \geqslant 0$. If the initial mass of drug in the gastrointestinal tract and body are $m_1(0) = 20.0$ mg and $m_2(0) = 0.0$ mg what are the drug levels $m_1(t)$ and $m_2(t)$ for $t \geqslant 0$? What are the maximum values for each of the drug levels; at what times do these maximums occur? Sketch the corresponding trajectory in the phase plane.

(b) There is initially no drug in the gastrointestinal tract and body, $m_1(0) = 0.0$ mg and $m_2(0) = 0.0$ mg, and the drug ingestion rate is given by

$$r(t) = Re^{-at}, \quad 0 \leqslant t < 120.0 \text{ min},$$

$$r(t) = Re^{-a(t-120)}, \quad t \geqslant 120.0 \text{ min},$$

where $R = 2.0$ mg/min and $a = 0.2 \; 1/\text{min}$. What are the drug levels $m_1(t)$ and $m_2(t)$ for $t \geqslant 0$? What are the maximum levels for each of the drug levels; at what times do these maximums occur? What is the total amount of drug ingested?

3-2. Consider the mixing of a salt solution as described in Case Study 3-2 with the parameter values

$$Q_1 = 10.0 \text{ gal/min}, \quad L_1 = 5.0 \text{ gal},$$
$$Q_2 = 5.0 \text{ gal/min}, \quad L_2 = 10.0 \text{ gal}.$$

(a) What is the transfer function?

(b) What is the impulse response function?

(c) Assume the input concentration $C(t) = 0$ for $t \geqslant 0$. If the initial amounts of salt in the two tanks are $S_1(0) = 5.0$ lb and $S_2(0) = 0.0$ lb, what are $S_1(t)$ and $S_2(t)$ for $t \geqslant 0$? What are the maximum amounts of salt in each tank; at what times do these maximums occur? Sketch the trajectory in the phase plane.

(d) If initially there is no salt in either tank, $S_1(0) = 0.0$ lb and $S_2(0) = 0.0$ lb, and the input concentration is

$$C(t) = 0.1 \text{ lb/gal}, \quad 0 \leqslant t < 10.0 \text{ min},$$

$$C(t) = 0.0 \text{ lb/gal}, \quad t \geqslant 10.0 \text{ min},$$

what are $S_1(t)$ and $S_2(t)$ for $t \geqslant 0$? What are the maximum amounts of salt in each tank; at what times do these maximums occur? What is the total amount of salt added to the tanks?

3-3.   Consider the Case Study 3-2 with the parameter values

$$L_1 = 10 \text{ gal}, \quad L_2 = 5 \text{ gal}, \quad Q_1 - Q_2 = 2 \text{ gal}/\text{min}.$$

What should the individual pumping rates Q_1 and Q_2 be so that the concentration of salt in the outflow line tends to the concentration of salt in the inflow line as rapidly as possible? Assume pumps are irreversible, i.e., $Q_1 > 0, Q_2 > 0$.

Does your answer depend on the initial amounts of salt in the tanks? How?

Does your answer depend on the level of salt concentration in the inflow line? How?

3-4.   Consider Case Study 3-4 for a bridged-T-filter. Assume the parameter values in the model are

$$C = 2.0 \times 10^{-6} \text{ farad}, \quad m = 1,$$
$$R = 100.0 \text{ ohm}, \quad n = 100.$$

(a) What is the transfer function?
(b) What is the damping ratio?
(c) What is the notch frequency of the circuit?
(d) What is the zero input response if initially the capacitor charges are $Q_1(0) = 10^{-6}$ coulombs, $Q_2(0) = 10^{-6}$ coulombs?
(e) What is the steady state response of the circuit if the input voltage is

$$e_i(t) = 1.0 \cos 300t + 2.0 \sin 100t \text{ volts}$$

for $t \geqslant 0$?

3-5.   Consider the electrical circuit

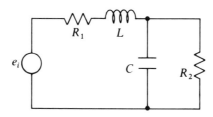

containing two resistors, a capacitor and an inductor. The input is an ideal voltage source applied to the indicated terminals; the output is the voltage drop across the second resistor.

Can the voltages across each of the two resistors be chosen as state variables? If so, what are the state equations and the transfer function? If not, discuss the difficulty.

3-6.   An electrical circuit is known to be linear; its input and output are both voltages. If the input voltage to the circuit is constant,

$$u(t) = 0.5 \text{ volt}, \quad t \geqslant 0,$$

then the zero state output voltage is measured to be

$$y(t) = 0.5[1 + e^{-20t}] - e^{-10t} \text{ volts for } t \geqslant 0 \text{ in seconds}.$$

(a) What is the transfer function?
(b) Is the circuit stable? Why?
(c) Give an all integrator block diagram for the circuit.
(d) What are state equations for the circuit?
(e) Give a schematic circuit realization using resistors, capacitors, inductors.

3-7. A linear RC circuit, with two capacitors, is described by state equations using two voltage state variables x_1 and x_2. The zero state response to a constant voltage input

$$u(t) = 2 \text{ volts}, \quad t \geqslant 0$$

is characterized by the trajectory indicated.

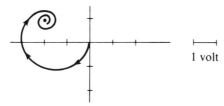

(a) Does the damping ratio satisfy

$$\zeta < 0, \quad 0 < \zeta < 1 \quad \text{or} \quad \zeta > 1?$$

(b) Sketch approximately the time response x_1 versus t and x_2 versus t, corresponding to the above trajectory.
(c) Sketch approximately the trajectory corresponding to the input $u(t) = 0$, $t \geqslant 0$ and the initial state $x_1(0) = 1$ volt $x_2(0) = -2$ volts.

3-8. A linear voltage amplifier is known to be stable with frequency response plots as indicated below.

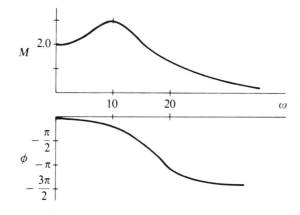

(a) Can the amplifier be first order; can it be second order; can it be third order?

(b) Are all of the characteristic zeros real? Why?

(c) What is the steady state response of the amplifier to a constant input voltage

$$u(t) = 2.0 \text{ volts} \quad \text{for } t \geqslant 0?$$

(d) What is the steady state response of the amplifier to a periodic input voltage

$$u(t) = 1.0 \cos 10t + 2.0 \sin 25t \text{ volts}, \quad t \geqslant 0?$$

3-9.   The motion of a rotated antenna is described by

$$I\frac{d^2\theta}{dt^2} + \mu\frac{d\theta}{dt} = T,$$

where θ is the angular displacement of the antenna from some reference. The moment of inertia of the antenna is I and μ is the antenna friction constant. A control torque T is applied to the antenna by a motor to control the direction of the antenna; this torque is given by

$$T = k(\theta_d - \theta),$$

where k is a motor constant and θ_d, the desired angle of the antenna, is considered to be the input; the actual displacement θ of the antenna is the output.

(a) Give a block diagram which exposes the feedback structure explicitly; indicate the signals θ_d, T and θ.

(b) Choose state variables and write state equations.

(c) What is the transfer function for the controlled system?

(d) For what values of k is the controlled system stable?

(e) What is the damping ratio?

3-10. A hot metal block is immersed in a cooling liquid bath. Assume the temperature is uniform throughout the block and the temperature of the liquid is uniform. The temperature of the air surrounding the liquid bath is also assumed to be uniform. The temperature of the air is the input; the temperature of the metal block is the output.

(a) Based on Newton's Law of cooling derive a state model for the system.

(b) What is the transfer function for the system?

(c) What is the damping ratio?

(d) If the temperature of the air is constant, show that the temperature of the metal block always tends to the temperature of the air no matter what the initial temperature of the metal block and the liquid bath.

3-11. The two dimensional motion of a boat moving in a river of width L is considered. Assume a fixed coordinate system so that the coordinate values x_1 and x_2 represent the cross stream and down stream positions of the boat, respectively. It is assumed that there is a constant downstream river current W. Thus the motion of the boat is described by the equations

$$\frac{dx_1}{dt} = V \cos \alpha,$$

$$\frac{dx_2}{dt} = V \sin \alpha + W,$$

where V is the constant velocity of the boat relative to the river and α is the heading angle of the boat. Assume the following parameter values

$$W = 2.0 \text{ miles/hr}, \quad V = 10.0 \text{ miles/hr}, \quad L = 1.0 \text{ mile}$$

and assume that the initial location of the boat is

$$x_1(0) = 0, \quad x_2(0) = 0.$$

Consider a constant heading angle $\alpha(t) = 10.0$ degrees for $t \geqslant 0$. How long does it take for the boat to cross the river? What is the path of the boat? What is the downstream location of the boat when the opposite shore is reached?

3-12. A container holds 1.0 kg of water at 20.0°C. A 0.5 kg mass of aluminum is added to the water; the initial temperature of the aluminum is 75.0°C. Assume that the container is insulated so that no heat is lost to the surroundings. The specific heat of water is 1.0; the specific heat of aluminum is 0.2. In 1.0 min the temperature of the water is observed to increase by 2.0°C. Develop a state model describing the temperature of the water and the temperature of the aluminum. Using the state model what is the temperature of the water and the temperature of the aluminum for $t \geqslant 0$?

3-13. An example of a lead-lag network is shown below.

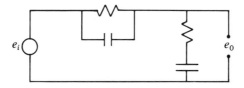

Assume each resistance is 10,000 ohms and each capacitance is 10^{-6} farad. The input voltage is e_i and the output voltage is e_0, as indicated.

(a) Derive state equations.

(b) What is the transfer function of the system?

(c) If the input voltage is zero, what is the time response of the circuit if initially there is a 1.0 volt drop across each of the capacitors? Sketch the corresponding trajectory in the phase plane.

(d) If the input is $e_i(t) = 1.0 \cos 100t$ volts for $t \geqslant 0$ what is the steady state output response of the circuit?

3-14. Consider the electrical filter

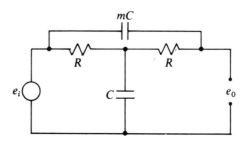

with the parameter values $C = 10^{-6}$ farads, $R = 10^4$ ohm, $m = 10$ and with input voltage e_i and output voltage e_0.

(a) Choose state variables and write state equations.

(b) What is the transfer function?

(c) What are the poles of the transfer function?

(d) If the input voltage is constant

$$e_i(t) = 10 \text{ volts}, \quad t \geqslant 0,$$

what is the steady state output voltage?

3-15. An electrical circuit is indicated below. The inductance is 1.0 henry, the capacitance is 0.5 farad and the resistance is 3.0 ohms. The input is the ideal current source as indicated and the output is the current through the capacitor.

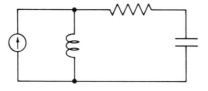

(a) Choose state variables and write state equations for the circuit.

(b) What is the transfer function for the circuit?

(c) What is the damping ratio?

(d) Sketch the frequency response plots, i.e., the magnitude and phase of the steady state output as a function of the input frequency.

3-16. An iron core audio transformer is indicated schematically below.

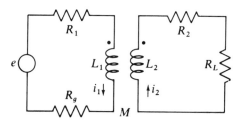

The resistances R_g, R_1, R_2, R_L and inductances L_1, L_2 and mutual inductance M are as indicated. The input is the ideal voltage source e; the output is the voltage drop across the resistance R_L; The circuit equations, in terms of loop currents i_1 and i_2 as indicated, are

$$e = (R_g + R_1)i_1 + L_1 \frac{di_1}{dt} - M \frac{di_2}{dt},$$

$$0 = - M \frac{di_1}{dt} + (R_L + R_2)i_2 + L_2 \frac{di_2}{dt}.$$

Assume the parameter values are

$$R_1 + R_g = 3.3 \text{ ohm}, \quad R_2 = 0.75 \text{ ohm}, \quad R_L = 2.0 \text{ ohm},$$

$$L_1 = 0.1 \text{ henry}, \quad L_2 = 0.01 \text{ henry}, \quad M = 0.025 \text{ henry}.$$

(a) Choose state variables and write state equations.
(b) What is the transfer function for the transformer?
(c) Is the circuit stable; what is the damping ratio?
(d) If the voltage input is

$$e(t) = 10.0 \cos 360t \text{ volts} \quad \text{for } t \geqslant 0$$

what is the steady state output voltage of the transformer?

3-17. A simple model for the vertical ascent of a deep sea diver is now suggested. Let h denote the depth of the diver below sea level and let p denote the internal pressure in the diver's internal organs. Assuming that the input v is the vertical velocity at which the diver is being raised then

$$\frac{dh}{dt} = v.$$

The rate of change of the internal pressure p is assumed to be proportional to the pressure difference between the diver's internal pressure and the local pressure of the water surrounding the diver; thus

$$\frac{dp}{dt} = k(wh - p),$$

where w is the weight density of water and k is a constant characterizing the diver's response to external pressure changes. The output of interest is the diver's internal pressure.

(a) Choose state variables and write state equations.
(b) What is the transfer function?
(c) Is the model stable?
(d) What are conditions for equilibrium?
(e) If the diver is raised at a constant velocity, $v(t) = V$ for $t \geqslant 0$, and initially the diver's depth is $h(0) = H$ and his internal pressure is $p(0) = P$, what is his internal pressure as a function of time thereafter?

3-18. As discussed in Case Study 3-3 a DC motor, with constant armature current, can be modelled by the equations

$$- v_f + R_f i_f + L_f \frac{di_f}{dt} = 0,$$

$$J \frac{d\omega}{dt} = - c\omega + BI_a i_f.$$

Suppose that a voltage amplifier and a velocity sensor are connected to the DC motor so that the field voltage v_f is generated according to the feedback equation

$$v_f = M(\omega_r - \omega),$$

where ω_r is a reference angular velocity which is now the input variable; the constant M is an amplifier and scale parameter. The output is the angular velocity ω of the motor. Assume the following parameter values.

$R = 5.0$ ohm, $L = 1.0$ henry, $BI_a = 10.0$ (ft-lb)/amp,

$J = 5.0$ (ft-lb-sec)/rad, $c = 1.0$ (ft-lb-sec)/rad,

$M = 50.0$ (volt-sec)/rad.

(a) Choose state variables and write state equations.
(b) What is the transfer function of this feedback system?
(c) Is the motor stable; what is the damping ratio?
(d) What is the zero state response for $0 \leqslant t \leqslant 2.0$ sec if the reference input velocity is

$\omega_r(t) = 10.0$ rad/sec, $0 \leqslant t < 1.0$ sec,

$\omega_r(t) = -10.0$ rad/sec, 1.0 sec $\leqslant t < 2.0$ sec?

(e) What is the zero input response if initially $i_f(0) = 0.0$ amp and $\omega(0) = 20.0$ rad/sec?

Chapter IV

Higher Order Linear State Models

In this chapter physical processes are considered which require more than two state variables for their description. A few theoretical results are first indicated and then the theory is used to develop and analyze state models for several physical processes.

Some Systems Theory

Consider linear state equations of the form

$$\frac{dx_1}{dt} = a_{11}x_1 + \cdots + a_{1n}x_n + b_n u,$$

$$\vdots$$

$$\frac{dx_n}{dt} = a_{n1}x_1 + \cdots + a_{nn}x_n + b_n u,$$

$$y = c_1 x_1 + \cdots + c_n x_n.$$

Here u is the input variable, y is the output variable and $x_1, \ldots, x_n$ are n state variables, where $n > 2$. In the case where more than two state variables are required, a detailed analysis of the response properties of the system is usually difficult; calculations can become quite involved and the dependence of the output response on the initial state and the input function may be quite complicated. The objective in this section is to indicate certain general procedures rather than to give precise formulas.

First consider the procedure for determining the transfer function from the state equations. Assuming the initial states are all zero, the Laplace transforms of the state equations can be obtained to yield n linear algebraic equations in the transforms $X_1(s), \ldots, X_n(s), U(s)$ of $x_1(t), \ldots, x_n(t), u(t)$. Then these algebraic equations can be solved to yield the transform of each state variable as a function of the transform of the input function. The equation for the output can be used to express the transform of the output in terms of the transform of the input $Y(s) = G(s)U(s)$ where $G(s)$ is then the transfer function. The transfer function is necessarily the ratio of two polynomial functions

$$G(s) = \frac{n(s)}{d(s)}$$

with the degree of the denominator polynomial exceeding the degree of the numerator polynomial. If $n(s_z) = 0$ then s_z is a zero of the transfer function; if $d(s_p) = 0$ then s_p is a pole of the transfer function. If there are no common factors between $n(s)$ and $d(s)$ then $d(s)$ is necessarily a polynomial of degree n so that the state realization is necessarily minimal.

Although it is difficult to characterize the zero input response explicitly certain qualitative properties of the response can be determined from the zeros of the characteristic polynomial $d(s)$, i.e., the poles of $G(s)$. If all of the characteristic zeros, some of which may be complex, have real parts which are negative then the zero input response necessarily tends asymptotically to zero; the system is said to be stable. Unfortunately if $n > 2$ it is not an easy matter to compute the characteristic zeros by factoring the characteristic polynomial. But there are procedures for determining if the characteristic zeros have negative real parts without performing a factorization. One simple tabular approach is called the Routh-Hurwitz criteria; it is described in Appendix C.

The zero state response of the system is of course determined by the transfer function. The transform of the zero state output response is related to the transform of the input function by

$$Y(s) = G(s)U(s).$$

The inverse transform can be expressed in terms of a convolution integral to obtain the expression for the zero state response as

$$y(t) = \int_0^t g(t - \tau)u(\tau)\, d\tau,$$

where $g(t)$ is the inverse transform of the transfer function, called the impulse response function or the weighting function of the system.

If the input function is constant, say $u(t) = \bar{u}$ for $t \geqslant 0$, then the equilibrium states can be determined by solving the algebraic equations

$$0 = a_{11}x_{11} + \quad \cdots + a_{1n}x_n + b_1\bar{u},$$
$$\vdots \qquad\qquad \vdots$$
$$0 = a_{n1}x_1 + \quad \cdots + a_{nn}x_n + b_n\bar{u}$$

which guarantee that $dx_1/dt = \cdots = dx_n/dt = 0$.

The corresponding constant output is easily determined from the output equation. An alternate approach is to determine the zero state response to the input $u(t) = \bar{u}$ for $t \geqslant 0$ using the transfer function; the transform of the zero state response is

$$Y = G(s)\frac{\bar{u}}{s} \; ;$$

assuming the system is stable and $G(s)$ does not have a pole at $s = 0$

$$Y = G(0)\frac{\bar{u}}{s} + \frac{p(s)\bar{u}}{d(s)}$$

for some polynomial $p(s)$; so that the zero state response satisfies

$$y(t) \rightarrow G(0)\bar{u} \quad \text{as } t \rightarrow \infty.$$

In fact, if the system is stable then for any initial state

$$y(t) \rightarrow G(0)\bar{u} \quad \text{as } t \rightarrow \infty.$$

Now consider the response to a sinusoidal input function $u(t) = \bar{u} \cos \omega t$ for $t \geqslant 0$ for constant $\bar{u}$. The transform of the zero state response is

$$Y = G(s)\frac{\bar{u}s}{s^2 + \omega^2} = \frac{n(s)\bar{u}s}{d(s)(s^2 + \omega^2)}$$

which can be written as

$$Y = \frac{M\bar{u}(s \cos \phi - \omega \sin \phi)}{s^2 + \omega^2} + \frac{p(s)\bar{u}}{d(s)} \; ,$$

where $Me^{j\phi} = G(j\omega)$. Assuming the system is stable

$$y(t) \rightarrow M(\omega)\bar{u} \cos(\omega t + \phi(\omega)) \quad \text{as } t \rightarrow \infty.$$

The latter expression is referred to as the steady state response to the sinusoidal input function $u(t) = \bar{u} \cos \omega t$ for $t \geqslant 0$. Thus the frequency response function $G(j\omega)$ characterizes the steady state response to sinusoidal input functions. As previously the amplitude $M(\omega)$ and phase $\phi(\omega)$ when plotted as a function of the input frequency ω, usually on a log-log scale, are referred to as Bode plots for the system.

This brief description of the analysis of higher order linear state equations can only serve as an introduction to the subject; extensive treatments of the material, using matrix theory, can be found in the texts by Braun, Boyce and DiPrima, Luenberger, and Polak and Wong.

Case Study 4-1: Reduction of Limestone by Heating

In this example a common industrial process is considered, namely the reduction of limestone by heating into its two main products, calcium and magnesium oxides. A reaction vessel, maintained at a constant high temperature, is used to carry out the chemical reactions. In particular the limestone is assumed to consist of a fraction β of $CaCo_3$ and a fraction $1 - \beta$ of $MgCO_3$, where $0 < \beta < 1$. At sufficiently high temperature these compounds undergo the following first order irreversible chemical reactions

$$CaCO_3 \rightarrow CaO + CO_2,$$

$$MgCO_3 \rightarrow MgO + CO_2.$$

Assume that the rate at which limestone is added to the reaction vessel is the input variable u. Let x_1 denote the mass of $CaCO_3$, in moles, in the reaction vessel. Similarly, let x_2, x_3, x_4 denote the mass, in moles, of $MgCO_3$, CaO, MgO, respectively, in the reaction vessel. Assuming the rate at which each reaction proceeds is proportional to the mass of the reactant in the reaction vessel, then it follows that

$$\frac{dx_1}{dt} = -k_1 x_1 + \beta u,$$

$$\frac{dx_2}{dt} = -k_2 x_2 + (1 - \beta)u,$$

where k_1 is a rate constant for the first reaction and k_2 is a rate constant for the second reaction. Obviously, each mole of reactant which decomposes yields one mole of product plus one mole of carbon dioxide CO_2. Thus the rates of formation of the products are

$$\frac{dx_3}{dt} = k_1 x_1,$$

$$\frac{dx_4}{dt} = k_2 x_2.$$

The carbon dioxide does not affect the rates at which the reactions occur; since its production is not of interest it need not be included as a state variable. Thus the model for reduction of limestone is a fourth order linear state model. The objective is to determine how the input variable affects each of the four state variables. It is important to recognize the simple structure of the state equations in this case. The first state equation is uncoupled from the others, as is the second state equation. Further the last two state equations are particularly simple with derivatives depending only on the first two state variables. These observations make the subsequent analysis particularly simple.

First, consider a closed reaction vessel to which no limestone is being added although there is initially M moles of limestone in the reaction vessel. Thus

$$u(t) = 0, \quad t \geqslant 0,$$

and

$$x_1(0) = \beta M, \quad x_2(0) = (1 - \beta)M,$$
$$x_3(0) = 0, \quad x_4(0) = 0.$$

The first state equation is easily solved, by itself, to obtain

$$x_1(t) = \beta M e^{-k_1 t};$$

the second state equation is easily solved, by itself, to obtain

$$x_2(t) = (1 - \beta)M e^{-k_2 t}.$$

Using the expression for x_1 the third state equation can be integrated to obtain

$$x_3(t) = \beta M [1 - e^{-k_1 t}].$$

Integrating the fourth state equation yields

$$x_4(t) = (1 - \beta)M [1 - e^{-k_2 t}].$$

Thus $x_1(t) \to 0$ and $x_2(t) \to 0$ as $t \to \infty$ while $x_3(t) \to \beta M$ and $x_4(t) \to (1 - \beta)M$ as $t \to \infty$. That is, all of the limestone is eventually used up and converted into βM moles of CaO and $(1 - \beta)M$ moles of MgO.

Now consider the case where the reaction vessel is initially empty so that

$$x_1(0) = 0, \quad x_2(0) = 0, \quad x_3(0) = 0, \quad x_4(0) = 0.$$

Then the following four transfer functions are easily computed

$$\frac{X_1(s)}{U(s)} = \frac{\beta}{s + k_1},$$

$$\frac{X_2(s)}{U(s)} = \frac{(1 - \beta)}{s + k_2},$$

$$\frac{X_3(s)}{U(s)} = \frac{k_1 \beta}{s(s + k_1)},$$

$$\frac{X_3(s)}{U(s)} = \frac{k_2(1 - \beta)}{s(s + k_2)}.$$

These expressions could be used to compute the zero state responses to any given input function. Note that the characteristic polynomial here is the fourth degree polynomial

$$d(s) = s^2(s + k_1)(s + k_2).$$

Thus the double zero at $s = 0$ is characteristic of an unstable system. The consequences of this instability are illustrated shortly.

A simple case is where limestone is added to the reaction vessel at a constant rate $u(t) = R$, $t \geqslant 0$. Thus, $U(s) = R/s$ and the above expressions can be used to determine the time responses

$$x_1(t) = \frac{\beta R}{k_1} \left[1 - e^{-k_1 t} \right],$$

$$x_2(t) = \frac{(1 - \beta)R}{k_2} \left[1 - e^{-k_2 t} \right],$$

$$x_3(t) = \frac{\beta R}{k_1} \left[k_1 t - 1 + e^{-k_1 t} \right],$$

$$x_4(t) = \frac{(1 - \beta)R}{k_2} \left[k_2 t - 1 + e^{-k_2 t} \right].$$

Thus the masses of $CaCO_3$ and $MgCO_3$ satisfy

$$x_1(t) \rightarrow \frac{\beta R}{k_1}, \quad x_2(t) \rightarrow \frac{(1 - \beta)R}{k_2} \quad \text{as } t \rightarrow \infty,$$

and the masses of the products of the reactions, CaO and MgO, continue to increase indefinitely at the asymptotic rates of increase βR and $(1 - \beta)R$, respectively. Such an unbounded response for a bounded input is characteristic of an unstable system.

The case where there is some limestone initially in the reaction vessel and limestone is added to the reaction vessel can be studied by adding the zero input and the zero state responses previously determined.

Case Study 4-2: Vertical Ascent of a Deep Sea Diver

It is important to raise a deep sea diver from great depths in a very precise way to prevent occurrence of extreme pressure changes in the body tissue of the diver. Typically, the diver is connected to a cable which is raised or lowered by a winch mechanism. Thus, in practice, an external force is exerted on the diver as a means of raising or lowering the diver in the water. The important physiological variable for the diver is the average internal pressure in his body tissue. Also of interest is the difference between that tissue pressure and the local pressure on the diver in the water. The objective here is to develop a mathematical model which can be used to study the relationships between the external winch force on the diver and the pressure variables just described.

The forces on a deep sea diver are his effective underwater weight, a drag force, and the external cable force. In this study only vertical motion of the diver is considered. Let h denote the depth of the diver below sea level. Based on Newton's law

$$\frac{W}{g}\frac{d^2h}{dt^2} = (W - wV) - \mu\frac{dh}{dt} - f.$$

Here W is the weight of the diver at sea level, V is his volume, w is the weight density of water and g is the acceleration of gravity as usual. The effective underwater weight of the diver, accounting for buoyancy, is $W - wV$; the drag force on the diver is assumed to be proportional to his velocity as indicated with μ denoting a drag coefficient. The force f is the external cable force on the diver. Let p denote the internal body pressure of the diver, relative to atmospheric pressure at sea level. The rate of change of the diver's internal pressure is assumed to be proportional to the difference between his internal pressure and the pressure in the water surrounding him. The local underwater pressure, relative to atmospheric pressure, is equal to the weight density of water multiplied by the diver's depth below sea level, wh. Thus

$$\frac{dp}{dt} = K(wh - p),$$

where K is a constant characterizing the body tissue of the diver. The above two equations, derived from physical considerations, can be used to study the dynamic changes in the defined variables.

Even though our interest is in the effect of the external force f as an input variable it is convenient to define the input variable to be the difference of the winch force and the constant effective weight of the diver, in acceleration units, namely

$$u = \left(\frac{W - wV}{W}\right)g - \left(\frac{g}{W}\right)f.$$

It is clear that three state variables are required to describe the physical process; a natural choice is the depth of the diver $x_1 = h$, the velocity of the diver $x_2 = dh/dt$ and the diver's internal pressure $x_3 = p$, so that the state model is given by

$$\frac{dx_1}{dt} = x_2,$$

$$\frac{dx_2}{dt} = -Dx_2 + u,$$

$$\frac{dx_3}{dt} = K(wx_1 - x_3).$$

For simplicity, a new drag coefficient has been introduced where

$$D = \frac{\mu g}{W}.$$

This is a third order linear state model. The first two state equations do not depend on the third state variable; hence they are uncoupled from the third state equation.

Conditions for equilibrium are first examined. The conditions for equilibrium are that

$$\frac{dx_1}{dt} = 0, \quad \frac{dx_2}{dt} = 0, \quad \frac{dx_3}{dt} = 0;$$

These conditions are satisfied only if the input $u(t) = 0$, $t \geqslant 0$, or equivalently the force f is a constant given by

$$f(t) = W - wV, \quad t \geqslant 0.$$

In addition, the conditions

$$x_2 = 0, \quad x_3 = wx_1$$

must be satisfied. Thus the diver is in equilibrium only if the external force on the diver just balances his effective weight, if his vertical velocity is zero, and if his internal body pressure exactly equals the local underwater pressure. The diver can be in equilibrium at any depth below sea level so long as these conditions are satisfied.

In the following analysis it is assumed that the diver is initially in equilibrium at a depth H below sea level; thus initially

$$x_1(0) = H, \quad x_2(0) = 0, \quad x_3(0) = wH.$$

The diver would remain in equilibrium thereafter if $f(t) = W - wV$, that is if $u(t) = 0$, $t \geqslant 0$. Our interest, of course, is in the more general case where $f(t) > W - wV$, $t \geqslant 0$, so that the diver is being raised in the water. For the present, assume that $u(t)$, $t \geqslant 0$, is arbitrary. Taking the Laplace transform of the state equations obtain

$$sX_1 - H = X_2,$$
$$sX_2 - 0 = -DX_2 + U,$$
$$sX_3 - wH = K(wX_1 - X_3).$$

Solving for X_1, X_2, X_3 obtain

$$X_1 = \frac{H}{s} + \frac{U}{s(s+D)}$$

$$X_2 = \frac{U}{s+D}$$

$$X_3 = \frac{wH}{s} + \frac{KU}{s(s+D)(s+K)} .$$

Thus the characteristic polynomial is

$$d(s) = s(s+D)(s+K).$$

The characteristic zeros are $0, -D, -K$ so that the system is unstable. Returning to the previous expressions, the inverse transforms of X_1, X_2, X_3

can be expressed in terms of convolution integrals as follows

$$x_1(t) = H + \frac{1}{D} \int_0^t [1 - e^{-D(t-\tau)}] u(\tau) \, d\tau,$$

$$X_2(t) = \int_0^t e^{-D(t-\tau)} u(\tau) \, d\tau,$$

$$x_3(t) = wH + \frac{1}{D(D-k)} \int_0^t [(D-K) + Ke^{-D(t-\tau)}$$

$$- De^{-K(t-\tau)}] u(\tau) \, d\tau,$$

assuming that $K \neq D$. Thus, assuming the diver is initially in equilibrium as indicated, expressions have been derived for the time responses.

Now consider the case where the external force exerted on the diver is a constant, $f(t) = F$, $t \geq 0$, such that $F > (W - wV)$. Thus

$$u(t) = \left(\frac{W - wV}{W} \right) g - \left(\frac{g}{W} \right) F, \quad t \geq 0,$$

is a negative constant. The time responses can be determined using the above expressions. Substituting for the value of $u(t)$, $t \geq 0$, and for the parameter D, the depth of the diver is given by

$$x_1(t) = H - \frac{(F - W + wV)}{\mu} \left\{ t - \frac{W}{\mu g} \left[1 - \exp\left(-\frac{\mu g t}{W} \right) \right] \right\}$$

and the internal pressure of the diver is given by

$$x_3(t) = wH - \frac{w(F - W + wV)}{\mu} \left\{ t - \frac{KW^2}{(\mu g)(KW - \mu g)} \left[1 - \exp\left(\frac{-\mu g t}{W} \right) \right] \right.$$

$$+ \frac{\mu g}{K(KW - \mu g)} \left[1 - \exp(-Kt) \right] \right\}.$$

The time T at which the diver reaches sea level could be obtained by solving the transcendental equation

$$x_1(T) = 0.$$

The corresponding internal pressure of the diver could be computed as $x_3(T)$ from the above expression. Typical time responses for these variables are illustrated in Figure 4-2-1.

Another variable of importance is the difference between the internal pressure of the diver and the local underwater pressure. This difference is

$$q(t) = x_3(t) - wx_1(t).$$

Clearly, $0 < q(t) < x_3(t)$, $t \geq 0$, and the time response for $q(t)$, $t \geq 0$, has the general form of $x_3(t)$, $t \geq 0$, as indicated above.

Only the case of a constant external force on the diver has been examined. Other input functions are certainly of interest. For example, the case where the diver is raised for a period of time and then allowed to

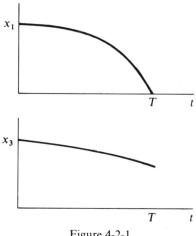

Figure 4-2-1

remain at a constant depth for a time period so that his internal pressure can adjust to the local underwater pressure is a situation of importance. Such manuevers could be analyzed using the model which has been developed.

A simple computer simulation is carried out for this case study example; the simulation program and results are given in Appendix E.

Case Study 4-3: A Speedometer Mechanism

One common type of speedometer mechanism, such as is used in automobiles, is shown pictorially in Figure 4-3-1. The underlying principle of the instrument is to develop a torque which is proportional to the rotational speed of an input shaft, i.e., the rotational speed of the wheels of the automobile. This torque in turn deflects an indicator needle through an angle proportional to the torque. Transmission of motion from the input shaft to the indicator needle is through a reaction damper element; this is a device filled with a viscous fluid which transmits a torque which is proportional to the difference in angular velocities of the input damper element and the reaction damper element.

As indicated in Figure 4-3-1 θ_1 is the angular displacement, from a fixed reference, of the input damper element; θ_2 is the angular displacement, from a fixed reference, of the reaction damper element; θ_2 is also the displacement of the speedometer indicator needle. The input θ_i is the angular displacement of the input shaft, i.e., the wheels of the automobile; the displacement θ_2 of the indicator needle is the output of the system. The constants J_1 and J_2 are moments of inertia of the two damper elements and their connecting shafts; k_1 and k_2 are the torsional spring constants of the

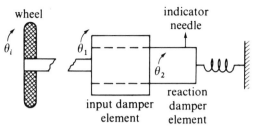

Figure 4-3-1

transmission shaft and the indicator spring; μ is a proportionality factor for the reaction damper element.

Using Newton's laws the moment of inertia of the input damper element multiplied by its angular acceleration is equal to the sum of all torques acting on the input damper element, that is

$$J_1 \frac{d^2\theta_1}{dt^2} = -k_1(\theta_1 - \theta_i) - \mu\left(\frac{d\theta_1}{dt} - \frac{d\theta_2}{dt}\right).$$

Similarly for the reaction damper element

$$J_2 \frac{d^2\theta_2}{dt^2} = \mu\left(\frac{d\theta_1}{dt} - \frac{d\theta_2}{dt}\right) - k_2\theta_2.$$

These are the physical equations which describe the motion of the elements of the speedometer mechanism; our interest is to determine how the rotational speed of the input shaft affects the displacement of the indicator needle.

In this case the physical equations are each of second order so it is expected that four state variables are required. One choice of state variables consists of the two angular position variables $x_1 = \theta_1, x_3 = \theta_3$ and the two angular velocity variables $x_2 = d\theta_1/dt, x_4 = d\theta_2/dt$. The input variable u is the angular position θ_i of the tire; the output variable y is the angular displacement θ_2.

Thus the state model is given by

$$\frac{dx_1}{dt} = x_2,$$

$$\frac{dx_2}{dt} = -\frac{k_1}{J_1}(x_1 - u) - \frac{\mu}{J_1}(x_2 - x_4),$$

$$\frac{dx_3}{dt} = x_4,$$

$$\frac{dx_4}{dt} = \frac{\mu}{J_2}(x_2 - x_4) - \frac{k_2}{J_2}x_3,$$

$$y = x_3.$$

The transfer function is first determined; it can be found by taking the transform of the state equations, assuming the initial state is zero, or it can be found by taking the transform of the physical equations. This latter approach is somewhat more direct but in either case the transfer function is given by

$$G(s) = \frac{k_1 \mu s}{(J_1 s^2 + \mu s + k_1)(J_2 s^2 + \mu s + k_2) - \mu^2 s^2},$$

and the characteristic polynomial is

$$d(s) = (J_1 s^2 + \mu s + k_1)(J_2 s^2 + \mu s + k_2) - \mu^2 s^2.$$

This characteristic polynomial is difficult to factor, but the location of its characteristic zeros can be determined using the Routh-Hurwitz criteria. The Routh array, using results in Appendix C, is

$$
\begin{array}{ccc}
J_1 J_2 & (J_1 k_2 + J_2 k_1) & k_1 k_2 \\[2mm]
\mu(J_1 + J_2) & \mu(k_1 + k_2) & 0 \\[2mm]
\dfrac{J_1^2 k_2 + J_2^2 k_1}{J_1 + J_2} & k_1 k_2 & 0 \\[4mm]
\dfrac{\mu(J_1 k_2 - J_2 k_1)^2}{J_1^2 k_1 + J_2^2 k_1} & 0 & 0 \\[4mm]
k_1 k_2 & 0 & 0
\end{array}
$$

All coefficients of the characteristic polynomial are positive assuming that $J_1 k_2 - J_2 k_1 \neq 0$; thus the characteristic zeros necessarily all have negative real parts so that the system is stable. Thus, for any initial state the zero input response satisfies $y(t) \to 0$ as $t \to \infty$.

Now consider the zero state response to certain input functions of interest. First consider the case where the automobile is moving at a constant speed so that $u(t) = \omega t$ for $t \geq 0$ where ω is the constant angular velocity of the wheels. Then the transform of the zero state response is

$$Y = \left(\frac{k_1 \mu s}{d(s)}\right)\left(\frac{\omega}{s^2}\right) = \frac{k_1 \mu \omega}{s d(s)}$$

which can be written as

$$Y = \frac{\mu \omega}{k_2} \frac{1}{s} + \frac{p(s)}{d(s)},$$

when $p(s)$ is a polynomial of degree three or less. Thus the zero state response to $u(t) = \omega t$ satisfies

$$y(t) \to \frac{\mu \omega}{k_2} \quad \text{as } t \to \infty.$$

In fact, for any initial state the response to $u(t) = \omega t$ satisfies

$$y(t) \to \frac{\mu \omega}{k_2} \quad \text{as } t \to \infty.$$

That is, the displacement of the indicator needle tends to a constant value which is directly proportional to the speed of the automobile. Note that the calibration of the indicator dial depends only on the properties of the reaction damper and the indicator spring as characterized by the parameters μ and k_2.

Now consider the case where the automobile is moving at a constant acceleration so that $u(t) = \alpha t^2/2$ where α is the constant angular acceleration of the wheels. Then the transform of the zero state response is

$$Y = \left(\frac{k_1 \mu s}{d(s)}\right)\left(\frac{\alpha}{s^3}\right) = \frac{k_1 \mu \alpha}{s^2 d(s)}$$

which can be written as

$$Y = \frac{\mu \alpha}{k_2} \frac{1}{s^2} - \frac{\alpha \mu^2 (k_1 + k_2)}{k_1 k_2^2} \frac{1}{s} + \frac{p(s)}{d(s)}$$

for some polynomial $p(s)$.

Thus the zero state response to $u(t) = \alpha t^2/2$ satisfies

$$y(t) \to \left(\frac{\mu \alpha}{k_2}\right)t - \frac{\alpha \mu^2 (k_1 + k_2)}{k_1 k_2^2} \qquad \text{as } t \to \infty.$$

Since the actual angular velocity of the wheels is given by αt for $t \geqslant 0$, if the speedometer is calibrated as before for constant speed input then for constant acceleration input the indicator needle indicates a speed which is smaller than the actual speed by the constant factor

$$\frac{\alpha \mu^2 (k_1 + k_2)}{k_1 k_2^2}.$$

Of course the accuracy of the speedometer is most important when the automobile is moving at constant speed.

Case Study 4-4: Automobile Suspension System

One way of analyzing the vertical motion of an automobile suspension system is to consider the suspension system for one wheel in isolation. To keep the analysis as simple as possible consider a model for the suspension system as shown in Figure 4-4-1.

The mass m_1 represents one quarter of the mass of the automobile frame, m_2 represents the effective mass of the wheel. The constant k_1 is the stiffness of the suspension spring, k_2 is the stiffness constant for the tire and μ is the damping constant of the shock absorber. Assuming there is some reference level for the ground the car and wheel assembly define an equilibrium level. Then as measured from this equilibrium level, q_1 denotes the vertical displacement of the automobile frame, i.e., of the mass m_1, and q_2 denotes the vertical displacement of the wheel, i.e., of the mass m_2. The

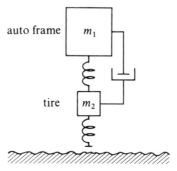

Figure 4-4-1

input is assumed to be the variable q_i which is the vertical displacement of the bottom of the tire due to the road irregularities, as measured from the ground reference level. The bottom of the tire is assumed to remain in contact with the ground. Using the simple spring-mass-damper model shown the mathematical equations which describe the motion of the suspension system are obtained by applying Newton's law to each mass in the system. This leads to the equations

$$m_1 \frac{d^2 q_1}{dt^2} = -k_1(q_1 - q_2) - \mu\left(\frac{dq_1}{dt} - \frac{dq_2}{dt}\right),$$

$$m_2 \frac{d^2 q_2}{dt^2} = -k_2(q_2 - q_i) + k_1(q_1 - q_2)$$

$$+ \mu\left(\frac{dq_1}{dt} - \frac{dq_2}{dt}\right).$$

It is natural to choose the displacements and velocities of each mass as the state variables, i.e., $x_1 = q_1, x_2 = dq_1/dt, x_3 = q_2, x_4 = dq_2/dt$. The input variable u is the displacement of the road surface q_i; the output variable y is the displacement of the automobile frame q_1.

The resulting state model is

$$\frac{dx_1}{dt} = x_2,$$

$$\frac{dx_2}{dt} = -\frac{k_1}{m_1}(x_1 - x_3) - \frac{\mu}{m_1}(x_2 - x_4),$$

$$\frac{dx_3}{dt} = x_4,$$

$$\frac{dx_4}{dt} = -\frac{k_2}{m_2}(x_3 - u) + \frac{k_1}{m_2}(x_1 - x_3) + \frac{\mu}{m_2}(x_2 - x_4),$$

$$y = x_1.$$

Thus the model for the automobile suspension is fourth order linear state equations. It is exceedingly difficult to determine exact analytical expressions for the response functions in terms of the given parameters. However, it is relatively easy to determine certain qualitative features of the response properties. The transfer function of the system can be determined by taking the transform of the four state equations, assuming the initial state is zero; then by solving the resulting four algebraic equations an expression for the transform of the output can be obtained as a function of the transform of the input.

After these involved calculations the obtained transfer function is given by

$$G(s) = \frac{k_2(\mu s + k_1)}{(m_1 s^2 + \mu s + k_1)(m_2 s^2 + \mu s + k_1 + k_2) - (\mu s + k_1)^2}.$$

Thus the characteristic polynomial is

$$d(s) = (m_1 s^2 + \mu s + k_1)(m_2 s^2 + \mu s + k_1 + k_2) - (\mu s + k_1)^2.$$

Since the characteristic polynomial is of degree four it is not possible to determine explicitly the characteristic zeros. But the stability can be checked using the Routh-Hurwitz criteria; each coefficient in the characteristic polynomial is positive and the Routh-Hurwitz array is

$m_1 m_2$	$(m_1 + m_2)k_1 + m_1 k_2$	$k_1 k_2$
$\mu(m_1 + m_2)$	μk_2	0
$\dfrac{k_1(m_1 + m_2)^2 + k_2 m_1^2}{(m_1 + m_2)}$	$k_1 k_2$	0
$\dfrac{\mu m_1^2 k_2^2}{k_1(m_1 + m_2)^2 + k_2 m_1^2}$	0	0
$k_1 k_2$	0	0

Since each term in the first column is positive it follows that all characteristic zeros have negative real part and the system is stable.

Consequently, for any initial state the zero input response always tends to zero. This fact certainly agrees with our expectation for the responses of an automobile suspension system.

Now consider the response of the automobile suspension to a step input, e.g., as characterized by the automobile running over a curb. Suppose that $u(t) = \bar{u}$ for $t \geq 0$; then the transform of the zero state response is given by

$$Y = \left[\frac{k_2(\mu s + k_1)}{d(s)} \right]\left(\frac{\bar{u}}{s} \right)$$

which can be written as

$$Y = \frac{\bar{u}}{s} + \frac{p(s)}{d(s)},$$

where $p(s)$ is a polynomial of degree three or less. Thus, since the system is stable it follows that the zero state response to step input satisfies

$$y(t) \to \bar{u} \quad \text{as } t \to \infty;$$

in fact, for any initial state the response to a step input satisfies

$$y(t) \to \bar{u} \quad \text{as } t \to \infty.$$

Now consider the response of an automobile suspension system to a sinusoidal input function, e.g., as characterized by the automobile running over a uniformly bumpy road. Suppose that $u(t) = \bar{u} \cos \omega t$ for $t \geq 0$. The transform of the zero state response is

$$Y = \left[\frac{k_2(\mu s + k_1)}{d(s)} \right] \left(\frac{\bar{u}s}{s^2 + \omega^2} \right).$$

It is possible to determine the inverse transform of this expression, although it is quite involved. It is much easier to recognize that, since the system is stable, the zero state response satisfies

$$y(t) \to M\bar{u} \cos(\omega t + \phi) \quad \text{as } t \to \infty,$$

where $M(\omega)e^{j\phi(\omega)} = G(j\omega)$. Of course, for any initial state

$$y(t) \to M\bar{u} \cos(\omega t + \phi) \quad \text{as } t \to \infty.$$

The frequency response functions $M(\omega)$ and $\phi(\omega)$ are determined from

$$M(\omega)e^{j\phi(\omega)} = \frac{k_2(j\mu\omega + k_1)}{d(j\omega)}.$$

After some rather complicated complex algebra it can be determined that $M(\omega)$

$$= \left\{ \frac{k_2^2 \left[k_1^2 + \mu^2\omega^2 \right]}{\left[(k_1 - m_1\omega^2)(k_2 - m_2\omega^2) - k_1 m_1 \omega^2 \right]^2 + \mu^2\omega^2 \left[k_2 - (m_1 + m_2)\omega^2 \right]^2} \right\}^{1/2},$$

$\tan \phi(\omega)$

$$= \frac{-m_1 \mu\omega^3 (k_2 - m_2\omega^2)}{k_1 \left[(k_1 - m_1\omega^2)(k_2 - m_2\omega^2) - k_1 m_1 \omega^2 \right] + \mu^2\omega^2 \left[k_2 - (m_1 + m_2)\omega^2 \right]}.$$

It can be seen that the frequency response functions $M(\omega)$ and $\phi(\omega)$ have the following general properties: if $\omega = 0$ then $M(\omega) = 1$, $\phi(\omega) = 0$; for $0 < \omega < \sqrt{k_2/m_2}$, $-\pi < \phi(\omega) < 0$; if $\omega = \sqrt{k_2/m_2}$ then $M(\omega) = m_2/m_1$ and $\phi(\omega) = -\pi$; for $\omega > \sqrt{k_2/m_2}$, $-\pi < \phi(\omega) < 3\pi/2$; as $\omega \to \infty$, $M(\omega) \to 0$ and $\phi(\omega) \to -3\pi/2$. The Bode plot of these frequency response functions are shown, approximately, in Figure 4-4-2.

The Bode plots make it possible to infer the effects of the frequency ω, determined by the properties of the road and speed of the automobile, on

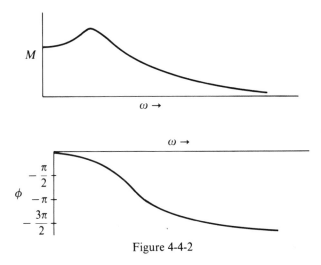

Figure 4-4-2

the displacement of the frame of the automobile. Such information is particularly valuable in evaluation or design of a suspension system.

Case Study 4-5: Magnetic Loudspeaker

Many technical and scientific applications require means for converting electrical energy into acoustical energy. Devices for converting energy from one form to another form are called transducers. An important device of this type is the common audio loudspeaker which converts electrical energy into acoustical energy. In the following a physical mechanism for realizing such energy conversion is considered. One widely used type of transducer, or loudspeaker, is illustrated by the schematic diagram in Figure 4-5-1.

A nonconducting coil form is attached to the loudspeaker cone; the periphery of the cone has a corrugated edge attached to a rigid outer support which is stationary with respect to the magnetic structure. The cone and the attached coil form are arranged to support a conducting coil which can move axially in the air gap of a permanent magnet. If there is a current in the coil then there is a magnetically induced force on the coil causing a resulting motion of the coil and loudspeaker cone. Sound is generated by

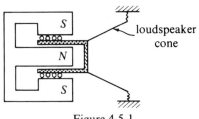

Figure 4-5-1

the action of the cone on the surrounding air. The elastic construction of
the cone serves to create a force on the coil opposing its motion and
tending to restore the coil to the center of the air gap. As the cone moves in
the air there is a viscous damping force, proportional to the velocity of the
cone, which also opposes the motion of the cone. The total mass of the
moving system is composed of the coil, the coil form and a portion of the
cone. The coil is an element of a simple circuit which includes a voltage
source; this voltage source is the mechanism by which electrical energy is
supplied to the loudspeaker. Our interest is to develop a mathematical
model which can be used to determine the response characteristics of the
loudspeaker.

The important variables here are the voltage source e, the current i in the
coil and the axial displacement z of the movable coil from its rest position
in the center of the air gap of the permanent magnet. The mathematical
description of the electrical circuit is given by

$$L\frac{di}{dt} + Ri + B\frac{dz}{dt} - e = 0.$$

Here L is the constant inductance of the coil, R is the total resistance of the
circuit, and $B(dz/dt)$ represents the voltage drop across the coil due to the
motion of the coil through the magnetic field. The dynamic motion of the
movable coil described according to Newton's law is

$$m\frac{d^2z}{dt^2} = -c\frac{dz}{dt} - kz + Bi.$$

Here m is the total mass of the movable coil and coil form; the first two
terms on the right hand side of the above equation are the viscous damping
and elastic forces on the cone (and hence on the coil); the force Bi
represents the magnetic force on the coil due to the current in the coil. The
constants c and k are viscous damping and elastic constants characteristic
of the loudspeaker. The electroacoustic coupling coefficient B depends on
the physical characteristics of the permanent magnet material and on the
construction of the movable coil; it is assumed that there is a uniform
magnetic field in the air gap of the magnet so that B is a constant.

A straightforward choice of state variables is the current through the
coil, the displacement and velocity of the coil; thus $x_1 = i$, $x_2 = z$, x_3
$= dz/dt$. The input variable u is the external voltage e applied to the
loudspeaker.

Thus the state equations for the loudspeaker are

$$\frac{dx_1}{dt} = -\frac{R}{L}x_1 - \frac{B}{L}x_3 + \frac{u}{L},$$

$$\frac{dx_2}{dt} = x_3,$$

$$\frac{dx_3}{dt} = \frac{B}{m}x_1 - \frac{k}{m}x_2 - \frac{c}{m}x_3$$

which constitute a third order linear state model.

Suppose that the initial state variables are zero; the Laplace transforms of the state equations are

$$sX_1 = -\frac{R}{L}X_1 - \frac{B}{L}X_3 + \frac{U}{L},$$

$$sX_2 = X_3,$$

$$sX_3 = \frac{B}{m}X_1 - \frac{k}{m}X_2 - \frac{c}{m}X_3.$$

Solving for X_1 and X_2 in terms of U obtain

$$X_1 = \frac{(s^2 + (c/m)s + (k/m))U/L}{d(s)},$$

$$X_2 = \frac{(BU/mL)}{d(s)}$$

where the characteristic polynomial is

$$d(s) = \left(s + \frac{R}{L}\right)\left(s^2 + \frac{c}{m}s + \frac{k}{m}\right) + \frac{B^2}{mL}s$$

or in expanded form

$$d(s) = s^3 + \left[\frac{c}{m} + \frac{R}{L}\right]s^2 + \left[\frac{cR}{mL} + \frac{k}{m} + \frac{B^2}{mL}\right]s + \frac{kR}{mL}.$$

This cubic polynomial cannot be explicitly factored but the conditions for stability of the loudspeaker can be determined using the Routh-Hurwitz criteria. All of the coefficients in the characteristic polynomial are positive. The Routh array is

1	$\dfrac{cR}{mL} + \dfrac{k}{m} + \dfrac{B^2}{mL}$
$\dfrac{c}{m} + \dfrac{R}{L}$	$\dfrac{kR}{mL}$
$\dfrac{cR}{mL} + \dfrac{B^2}{mL} + \dfrac{k}{m}\left(\dfrac{cL}{cL + Rm}\right)$	0
$\dfrac{kR}{mL}$	0

so that, since all coefficients in the left column are positive, the loudspeaker is always stable.

First, consider the relation between the external voltage input applied to the loudspeaker and the current in the coil. This transfer function, as determined, is

$$G_1(s) = \frac{(s^2 + (c/m)s + (k/m))/L}{d(s)}.$$

It is convenient to consider the reciprocal of $G_1(s)$ which is referred to as the input impedance transfer function of the loudspeaker; the input impedance is thus

$$\frac{1}{G_1(s)} = \frac{(B^2/m)s}{s^2 + (c/m)s + (k/m)} + Ls + R.$$

The frequency response of this input impedance can be written as

$$\frac{1}{G_1(j\omega)} = M_1 e^{j\phi_1}.$$

After carrying out the complex algebra obtain the frequency response amplitude function

$$M_1 = \left\{ \left[R + \frac{B^2(c/m)\omega^2}{\left((k/m) - \omega^2\right)^2 + (c^2/m^2)\omega^2} \right]^2 \right.$$

$$\left. + \left[L\omega + \frac{B^2\omega\left((k/m) - \omega^2\right)}{\left((k/m) - \omega^2\right)^2 + (c^2/m^2)\omega^2} \right]^2 \right\}^{1/2}.$$

The particular form of the frequency response curve depends importantly on the values of the parameters of the model; however, it is clear that if $\omega = 0$ then $M_1 = R$ and as $\omega \to \infty$, $M_1 \to L\omega$. A typical form for this frequency plot, the input impedance gain, is shown in Figure 4-5-2.

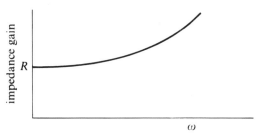

Figure 4-5-2

The frequency response of the input admittance gain of the transfer function $G_1(s)$ is easily obtained from the frequency response of the input impedance gain for $1/G_1(s)$. Clearly, if $\omega = 0$ then the admittance gain is $1/R$ and as $\omega \to \infty$ then the admittance gain decreases to zero. The general form of the input admittance gain, corresponding to the impedance gain shown in Figure 4-5-2, is shown in Figure 4-5-3.

Consideration of the phase angle relationships have been omitted in this analysis.

Next, consider the relation between the external voltage input applied to the loudspeaker and the velocity of the coil. This transfer function, as

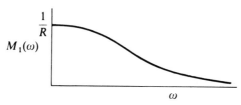

Figure 4-5-3

previously determined, is

$$G_3(s) = \frac{(B/mL)s}{d(s)}.$$

The frequency response of this transfer function can be written as

$$G_3(j\omega) = M_3 e^{j\phi_3}.$$

After carrying out the complex algebra it can be shown that the amplitude of the frequency response is given by

$$M_3 = (B/mL)\omega\left\{\left[(R/L)((k/m) - \omega^2) - (c/m)\omega^2\right]^2\right.$$

$$\left. + \omega^2\left[(B^2/mL) + (cR/mL) + ((k/m) - \omega^2)\right]^2\right\}^{-1/2}.$$

Observe that when $\omega = 0$ then $M_3 = 0$ and as $\omega \to \infty$ then $M_3 \to 0$. The general form of the frequency response curve is shown in Figure 4-5-4.

Again, the phase angle relationship for the transfer function $G_3(s)$ is not considered.

Now, consider the steady state responses of the loudspeaker in the case of a pure sinusoidal input function

$$u(t) = E \cos \omega t, \quad t \geq 0,$$

where E is the magnitude and ω is the frequency of the voltage input. Since the loudspeaker is stable there are steady state responses, and for any initial state the resulting responses tend asymptotically to the steady state responses. The steady state response of the current in the coil is

$$x_1(t) = M_1 E \cos(\omega t + \phi_1)$$

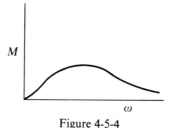

Figure 4-5-4

and the steady state response of the coil velocity is

$$x_3(t) = M_3 E \cos(\omega t + \phi_3).$$

The response of the loudspeaker to a general periodic input is important in evaluating the loudspeaker characteristics. Such input functions can be expanded in a Fourier cosine series of the form

$$u(t) = \sum_{i=1}^{\infty} E_i \cos(i\omega t + \xi_i), \quad t \geqslant 0,$$

for constants E_i, ω, ξ_i, $i = 1, 2, \ldots$. Using the superposition principle for linear systems the steady state response for the current in the coil is given by

$$x_1(t) = \sum_{i=1}^{\infty} M_1^i E_i \cos(i\omega t + \phi_1^i + \xi_i),$$

where

$$M_1^i = M_1(i\omega),$$

$$\phi_1^i = \phi_1(i\omega)$$

can be obtained graphically from the frequency response curves. Also the steady state response for the coil velocity is

$$x_3(t) = \sum_{i=1}^{\infty} M_3^i E_i \cos(i\omega t + \phi_3^i + \xi_i),$$

where

$$M_3^i = M_3(i\omega),$$

$$\phi_3^i = \phi_3(i\omega)$$

can be obtained graphically from the frequency response curves.

A further interpretation of the frequency response curves can be given. The electrical power supplied to the loudspeaker is proportional to the product of the applied voltage and current so that, in steady state, the magnitude of the electrical power supplied to the loudspeaker is

$$P_{\text{elec}} = \sum_{i=1}^{\infty} M_1^i E_i^2.$$

Hence, the electrical power supplied to the loudspeaker depends directly on the frequency response admittance gain $M_1(\omega)$.

The acoustical power produced by the loudspeaker is proportional to the product of the magnetic force on the coil and the velocity of the coil. In steady state this acoustic power is

$$P_{\text{acous}} = \sum_{i=1}^{\infty} BM_1^i M_3^i E_i^2.$$

Hence the acoustic power produced by the loudspeaker is directly related to the frequency response functions $M_1(\omega)$ and $M_3(\omega)$. The product $M_1 M_3$ is easily found from the individual frequency response functions. Note that at low frequencies, as $\omega \to 0$, the acoustical power produced by the loud-speaker tends to zero since M_3 has this property. This is an important limitation in the design of loudspeakers of the type described.

Exercises

4-1. Consider the model for the vertical ascent of a deep sea diver as developed in Case Study 4-2. Assume the parameter values in the model are

$$W = 200.0 \text{ lb}, \quad \mu = 5.0 \text{ (lb-sec)}/\text{ft},$$

$$K = 0.2 \text{ 1}/\text{sec}, \quad w = 62.4 \text{ lb}/\text{ft}^3, \quad V = 3.0 \text{ ft}^3.$$

Suppose that initially the diver is in equilibrium at a depth of 150.0 feet below sea level.

(a) Suppose that a constant winch force of 15.0 lb is applied to the diver. How long does it take to raise the diver to sea level? What is the maximum difference between the diver's internal pressure and the local underwater pressure?

(b) What should be the constant winch force with which the diver is raised if his body pressure is not to exceed the local underwater pressure by more than 5.0 lb/in²? How long does it take to raise the diver to sea level in such a case.

(c) A winch force, with sinusoidal time dependence, as shown below, is proposed for raising the diver.

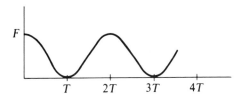

Choose the maximum force F and half period T so that the diver can be raised to sea level "as quickly as possible" without his internal body pressure exceeding the local underwater pressure by more than 5.0 lb/in². Are there any obvious advantages of such a scheme over that proposed in (b)?

4-2. Consider the speedometer mechanism as described in Case Study 4-3 with the following parameter values

$$J_1 = 100.0 \text{ (ft-lb-sec}^2)/\text{rad},$$

$$J_2 = 50.0 \text{ (ft-lb-sec}^2)/\text{rad},$$

$$k_1 = 50.0 \text{ (ft-lb)}/\text{rad}, \quad \mu = 50.0 \text{ (ft-lb-sec)}/\text{rad}.$$

(a) Choose k_2 so that the maximum deflection of the speedometer indicator needle, when the automobile is moving at 60.0 mph, is exactly 90°. Assume that the radius of the tires of the automobile is 1.0 ft.

(b) With the value of k_2 obtained in (a), what is the transfer function of the speedometer mechanism?

(c) Assume that k_2 has the value obtained in (a). Suppose that the automobile accelerates from rest with the velocity profile shown below. What is the response of the speedometer indicator needle for $0 \leqslant t \leqslant 20.0$ sec?

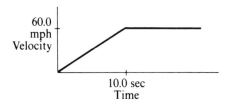

4-3. Consider the automobile suspension problem as described in Case Study 4-3 with the following parameter values.

$$m_1 g = 1300.0 \text{ lb},$$
$$m_2 g = 100.0 \text{ lb},$$
$$k_1 = 150.0 \text{ lb/in},$$
$$k_2 = 1100.0 \text{ lb/in},$$
$$\mu = 10.0 \text{ (lb-sec)/in}.$$

(a) What is the transfer function?

(b) Is the suspension system stable?

(c) Assume that the automobile is travelling along a smooth road at a constant forward speed of 30 mph when it hits a newly resurfaced section having the vertical profile as shown.

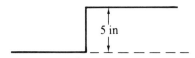

What is the vertical displacement of the automobile frame as a function of time?

(d) Assume that the automobile is travelling along a smooth road at a constant forward speed of 30 mph when it hits a bump in the road as shown

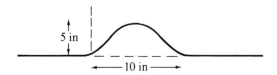

with the height of the bump given by

$$h(d) = 2.5\left[1 - \cos\frac{\pi d}{5}\right], \quad 0 \leqslant d \leqslant 10,$$

$$h(d) = 0, \quad d > 10,$$

where d is the distance from the leading point on the bump. Both h and d are measured in inches. What is the vertical displacement of the automobile frame as a function of time?

4-4. A tank holds three radioactive substances which decay in the following sequence

$$Ba \rightarrow La \rightarrow Ce.$$

The half life of Ba is 12.8 days and the half life of La is 40.2 hr. Assume that Ce is stable and does not decay. Assume that pure Ba is added to the tank at a certain rate, in grams per minute; this rate is the input variable. The output variable is the mass of Ce in the tank.

(a) What are the state equations?
(b) What is the transfer function from input to output?
(c) What is the impulse response function?
(d) Is the system stable? Why?
(e) Suppose that no Ba is added to the tank and the tank initially contains 5 grams of Ba and none of La or Ce. What is the subsequent amount of Ce in the tank?

4-5. Consider a reaction vessel for processing the chemical breakdown of the alcohol CH_3OH into an aldehyde CH_3O and an olefin CH_2, according to the reactions

$$CH_3OH \rightarrow CH_3O + H_2,$$

$$CH_3OH \rightarrow CH_2 + H_2O.$$

The reactions are assumed to occur concurrently, to be first order and irreversible and the reaction vessel is maintained at constant temperature.

The input is the rate at which the alcohol is added to the reaction vessel.

(a) Derive a state model for the process.
(b) If there is no alcohol added to the reaction vessel, what are the equilibrium states?
(c) If there is no alcohol added to the reaction vessel and the initial molar concentrations of alcohol, aldehyde and olefin are C_1, C_2, C_3 respectively, what are the subsequent molar concentrations of alcohol, aldehyde and olefin as the reactions proceed? Do the concentrations tend asymptotically to equilibrium values?
(d) If alcohol is added to the reaction vessel at a constant rate and initially there is no alcohol, aldehyde or olefin in the vessel what

are the subsequent molar concentrations of alcohol, aldehyde and olefin as the reaction proceeds?

4-6. Two rotating wheels with inertias J_1 and J_2 are coupled by a flexible shaft and a clutch mechanism as indicated in the figure below.

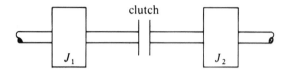

Let θ_1 and θ_2 denote the angular positions of the two wheels as measured from some reference. When the clutch is engaged the equations of motion of the two wheels are

$$J_1 \frac{d^2\theta_1}{dt^2} + K(\theta_1 - \theta_2) = 0,$$

$$J_2 \frac{d^2\theta_2}{dt^2} + K(\theta_2 - \theta_1) = 0,$$

where K is the stiffness coefficient of the flexible connecting shaft. All damping and friction torques are ignored.

(a) Choose state variables and write state equations which describe the motion of the two wheels.
(b) What is the characteristic polynomial for this mechanism? What are the characteristic zeros? These characteristic zeros define the natural frequencies of the mechanism. What is the physical significance of these natural frequencies?
(c) Assume that initially the second wheel is stationary and the first wheel is rotating at a constant angular velocity Ω. The clutch is then engaged instantaneously. Describe the resulting angular motion of the two wheels after the clutch is engaged.

4-7. Consider the first order irreversible chemical reaction

$$C_4H_{10} \rightarrow 2C_2H_4 + H_2$$

which is assumed to occur in a closed vessel at constant temperature and pressure. Develop a state model for the molar concentrations of C_4H_{10}, C_2H_4 and H_2 in the vessel. Assuming that the vessel is initially filled with C_4H_{10} only, what are the molar concentrations for $t \geqslant 0$. Assume that the molar concentration of C_4H_{10} is reduced by 50% in 30.0 sec.

4-8. Three tanks, each of volume 100 gal, are connected by flow lines as shown.

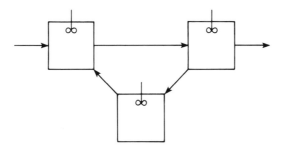

There is constant flow between the three tanks; the net inflow and outflow rates are 1 gal/min and the recirculation flow rate is 0.5 gal/min. Assume the tanks are completely filled with a salt solution which is kept at a uniform concentration in each tank by mixing. The input is the concentration of salt in the inflow line; the output is the concentration of salt in the outflow line.

(a) Choose state variables and write state equations.
(b) What is the transfer function of the system?
(c) If initially there is no salt in each of the three tanks and the concentration of salt in the inflow is constant 1 lb/gal for $t \geq 0$, what is the concentration of salt in the outflow for $t \geq 0$?

4-9. A projectile is fired at an angle $\theta = 45°$ from the horizontal, at the surface of the Earth. Let x, y denote the horizontal and vertical coordinates of the projectile, assuming a flat Earth. If only the gravitational force on the projectile is considered then the equations of motion of the projectile are

$$\frac{d^2x}{dt^2} = 0,$$

$$\frac{d^2y}{dt^2} = -g$$

where $g = 9.8$ meters/sec^2 is the constant acceleration of gravity. Assume the initial position and velocities of the projectile are

$$x(0) = 0, \quad \frac{dx}{dt}(0) = V\cos\theta,$$

$$y(0) = 0, \quad \frac{dy}{dt}(0) = V\sin\theta,$$

where $V = 250.0$ meters/sec is the initial velocity of the projectile.

(a) Choose state variables and write state equations.
(b) What is the maximum altitude of the projectile? How long does it take the projectile to reach this altitude?

(c) What is the range of the projectile? How long does it take the projectile to reach this range?

(d) Describe the path of the projectile.

4-10. A mathematical model for a magnetic microphone is given by

$$L\frac{dI}{dt} + (R_c + R)I + B\frac{dZ}{dt} = 0,$$

$$M\frac{d^2Z}{dt^2} + \mu\frac{dZ}{dt} + KZ - BI = F$$

where I is the current through the coil and Z is the displacement of the diaphragm element. The input is the force F on the diaphragm; the output is the voltage drop RI. The parameter values are

$$R_c = 9.0 \text{ ohm}, \quad R = 1.0 \text{ ohm},$$

$$L = 10^{-3} \text{ henry}, \ M = 0.01 \text{ kg},$$

$$\mu = 0.1 \ (\text{N-sec})/\text{meter},$$

$$K = 0.5 \text{ N/meter},$$

$$B = 0.3 \ (\text{volt-sec})/\text{meter}.$$

(a) Develop a state model.

(b) What is the transfer function of the microphone?

(c) Is the microphone stable?

(d) What is the zero state response to the input

$$F(t) = 0.05 \text{ N}, \quad 0 \leqslant t < 0.01,$$

$$F(t) = 0, \quad t \geqslant 0.01$$

for $t \geqslant 0$ in seconds?

(e) What is the steady state response to the input

$$F(t) = 0.005 \cos 50t \text{ N}$$

for $t \geqslant 0$ in seconds? Be careful to use correct units.

4-11. In this problem the two dimensional motion of a boat moving in a river of width L is considered. Assume there is a fixed coordinate system so that the location of the boat is given by the coordinate values x_1 and x_2 representing the cross stream and down stream position of the boat. It is assumed that there is a constant downstream river current of velocity W. The equations of motion of the boat are assumed to be given by

$$m\frac{d^2x_1}{dt^2} = T\cos\alpha - c,$$

$$m\frac{d^2x_2}{dt^2} = T\sin\alpha - c\left[\frac{dx_2}{dt} - W\right],$$

where m is the mass of the boat, T is the constant thrust of the boat's engine, α is the angle of the thrust vector with the x_1 coordinate direction and c is a drag coefficient for the boat. Suppose that the constant parameter values are

$$mg = 500.0 \text{ lbs}, \quad T = 25.0 \text{ lbs}, \quad W = 2.0 \text{ miles/hr},$$

$$c = 2.5 \text{ (lb-hr)/mile}, \quad L = 1.0 \text{ mile},$$

and that initially the position and velocity of the boat are

$$x_1(0) = 0, \quad x_2(0) = 0,$$

$$\frac{dx_1}{dt}(0) = 0, \quad \frac{dx_2}{dt}(0) = 0.$$

(a) Choose state variables and write state equations, assuming that the thrust angle α is the input variable.
(b) What is the path of the boat for a constant thrust angle $\alpha(t) = 10.0°$ for $t \geqslant 0$? How long does it take for the boat to cross the river? What is the path which the boat follows? What is the downstream location of the boat when the opposite shore is reached?

Chapter V

First Order Nonlinear State Models

In this and the succeeding chapters attention turns to examination of state models that are not linear. Consequently the techniques indicated previously which are valuable in analysis of linear state models are not of direct applicability here. In fact, it is not possible to develop a general theory of nonlinear state models; it is only possible to indicate some techniques and ideas which, for some nonlinear state models, may prove useful. Of course it is this characteristic of nonlinear state models which causes most difficulty; it is not possible to give general formulas or even procedures which are guaranteed to prove useful. As in the previous chapters some ideas and concepts are first presented; then several physical processes are examined where some of the concepts prove helpful.

Some Systems Theory

In this section a first order state model of the form

$$\frac{dx}{dt} = f(x, u),$$

$$y = g(x, u)$$

is examined; here u is the input variable, y is the output variable and x is a state variable.

The state model is defined by the two functions $f(x, u)$ and $g(x, u)$. Since analysis of the state model depends most importantly on the differential

96

equation

$$\frac{dx}{dt} = f(x, u)$$

most attention will be directed explicitly toward this equation.

Unlike a linear differential equation it is not generally possible to obtain an analytical solution expression for the state at an arbitrary time as an explicit function of the initial state and the input function. However, in some special cases it is possible to integrate the differential equation; in particular consider the special case of a constant input function $u(t) = \bar{u}$, $t \geq 0$. Then

$$\frac{dx}{dt} = f(x, \bar{u}).$$

Due to the special nature of this first order differential equation it can be approached using the method of separation of variables to obtain

$$\int_{x(0)}^{x(t)} \frac{dx}{f(x, \bar{u})} = \int_0^t d\omega.$$

In some cases, depending on the function $f(x, \bar{u})$, it is possible to express the definite integral above as an explicit function of $x(t)$, $x(0)$ and $\bar{u}$, say

$$\Phi(x(t), x(0), \bar{u}) = t.$$

Finally, it may be possible to use this implicit equation to determine $x(t)$ as an explicit function of t, $x(0)$ and $\bar{u}$. This is one possible way of attempting to integrate the state equation by separation of variables. Many other special integration techniques, e.g., using change of variables, are discussed in the texts by Braun and by Boyce and DiPrima.

Although it is not possible to solve the above differential equation for any initial state and any input function, it is usually rather easy to determine if there are constant solutions of the state equation. Assuming that the input is constant $u(t) = \bar{u}$ for $t \geq 0$, then if the condition

$$0 = f(\bar{x}, \bar{u})$$

holds for some $\bar{x}$ it follows that if $x(0) = \bar{x}$ then necessarily $x(t) = \bar{x}$ for all $t \geq 0$, i.e., $\bar{x}$ is an equilibrium state corresponding to the constant input $\bar{u}$; the corresponding output function is necessarily constant and is given by $y(t) = g(\bar{x}, \bar{u})$ for all $t \geq 0$.

One of the most important techniques in the analysis of nonlinear state equations is the approximation of the nonlinear equation by a suitable linear equation; there are several possible ways of defining such an approximation or linearization. One approach is the following. Suppose that $\bar{x}$ is an equilibrium state corresponding to the constant input function $\bar{u}$; then new variables v, w and z can be defined by

$$u = \bar{u} + v,$$

$$y = g(\bar{x}, \bar{u}) + w,$$

$$x = \bar{x} + z.$$

Thus obtain the equivalent nonlinear state model

$$\frac{dz}{dt} = f(\bar{x} + z, \bar{u} + v),$$

$$w = g(\bar{x} + z, \bar{u} + v) - g(\bar{x}, \bar{u}),$$

where v and w are new input and output variables and z is a new state variable. Clearly, if this latter state model can be analyzed then the results are directly applicable to the original state model. But this latter state model may be easily approximated by a linear state model. For if the input function u is near the constant function $\bar{u}$ and if the state function x is near the equilibrium state $\bar{x}$ then it is reasonable to approximate the function $f(x, u)$ and $g(x, u)$ by linear relations in x and u, as indicated in Appendix D, namely

$$f(\bar{x} + z, \bar{u} + v) \cong f(\bar{x}, \bar{u}) + \frac{\partial f}{\partial x}(\bar{x}, \bar{u})z + \frac{\partial f}{\partial u}(\bar{x}, \bar{u})v,$$

$$g(\bar{x} + z, \bar{u} + v) \cong g(\bar{x}, \bar{u}) + \frac{\partial g}{\partial x}(\bar{x}, \bar{u})z + \frac{\partial g}{\partial u}(\bar{x}, \bar{u})v$$

where the partial derivatives are evaluated at $\bar{x}$ and $\bar{u}$ as indicated. Substituting these approximations into the previous model and simplifying the result obtain

$$\frac{dz}{dt} = \left(\frac{\partial f}{\partial x}(\bar{x}, \bar{u}) \right)z + \left(\frac{\partial f}{\partial u}(\bar{x}, \bar{u}) \right)v,$$

$$w = \left(\frac{\partial g}{\partial x}(\bar{x}, \bar{u}) \right)z + \left(\frac{\partial g}{\partial u}(\bar{x}, \bar{u}) \right)v$$

which is a linear state model. Thus if the functions $u(t)$ and $x(t)$ are near the constant functions $\bar{u}$ and $\bar{x}$, the variables v and z are small in a certain sense; then the above linearized state model is a good approximation to the previous nonlinear state model. Consequently, under the stated assumptions for which the linearized state equations are valid, the linearized state model can be analyzed, as indicated in Chapter II, to characterize the response properties of the system. In particular, suppose that it is desired to determine the response corresponding to an initial state $x(0)$ near $\bar{x}$, and an input function u near $\bar{u}$. Then necessarily $z(0) = x(0) - \bar{x}$ and $v(t) = u(t) - \bar{u}$, $t \geqslant 0$; then, assuming that $z(t)$ is near zero it follows that the above linear state model can be used to determine $w(t)$ for $t \geqslant 0$ and consequently the output function $y(t) = g(\bar{x}, \bar{u}) + w(t)$ for $t \geqslant 0$. It is important to keep in mind that this linearization procedure is simply a rational way of approximating a nonlinear state model by a linear state model; and the approximation can be justified only under the stated assumptions.

Only a few ideas have been indicated which may prove useful in the analysis of first order nonlinear state models. Some specific physical processes are now examined to illustrate these, and other, systems concepts.

Case Study 5-1: Management of a Fisheries Resource

There are a number of applications of systems theory and modeling to problems in resource management. In this example a simple model of a renewable resource, namely a fisheries model, is developed to illustrate some of the important concepts in the development and use of nonlinear models. In the analysis several regulatory mechanisms are considered and the resulting effects on the resource are determined. Clearly, the model used in the development is a highly simplified one; however, important qualitative insight into the management of such a natural resource is obtained.

The particular situation considered is as follows. There is a single species of fish which is a renewable resource in the sense that the stock of fish is self-reproducing. The management problem is to harvest this resource in an efficient manner.

The dynamics of the fish population is first considered. If x is a measure of the size of the fisheries resource (for example the total weight of fish which can be harvested), then it is assumed that the rate of increase of the resource depends on the level of the existing resource according to the rate $Kx(M - x)$ for positive constants K and M. If u denotes the harvesting rate of the resource then the total range of change of the resource due to its own renewal and due to harvesting is

$$\frac{dx}{dt} = Kx(M - x) - u.$$

Throughout, it is assumed that $x \geqslant 0$ and $u \geqslant 0$. This simple model is the basis for the following analysis. Clearly the resource level x is the state variable and the harvesting rate u can be considered to be the input variable. Thus the above is a first order nonlinear dynamic model. In the following presentation, three different harvesting policies are examined as a way of obtaining insight into the resource management problem.

No Harvesting. First, the case where none of the resource is harvested is considered, i.e., $u(t) = 0$, $t \geqslant 0$, so that

$$\frac{dx}{dt} = Kx(M - x).$$

It is easy to plot the rate of change of the resource dx/dt as a function of the resource level x as in Figure 5-1-1.

Note that the fisheries resource is in equilibrium where $dx/dt = 0$, i.e.,

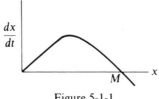

Figure 5-1-1

when $x = 0$ or when $x = M$. Further, the rate of change of the resource, dx/dt, is a maximum when $x = M/2$, with a maximum rate of $KM^2/4$. It is clear that if $0 < x < M$ then $dx/dt > 0$ so that the resource level is increasing, while if $x > M$ then $dx/dt < 0$ so that the resource level is decreasing. Thus for any nonzero initial state it follows that $x(t) \to M$ as $t \to \infty$, i.e., the resource level tends asymptotically to the constant level M. The equilibrium state $x = 0$ is said to be unstable; the equilibrium state $x = M$ is said to be stable. In this simple case the general qualitative features of the responses are easily obtained by elementary arguments. But as an illustration in this simple case the analytical solution of the state model can be determined. Using the separation of variables procedure obtain

$$\int \frac{1}{x(M - x)} \, dx = \int K \, dt + C,$$

where C is a constant of integration. Using the partial fraction relation

$$\frac{1}{x(M - x)} = \frac{1}{M}\left(\frac{1}{x} + \frac{1}{M - x}\right),$$

the left integral above is easily integrated in closed form to obtain

$$\frac{1}{M}\left[\ln(x) - \ln(M - x)\right] = Kt + C.$$

Assuming that the initial resource level is $x(0) = P$ where $P < M$ then necessarily the constant C is

$$C = \frac{1}{M}\left[\ln(P) - \ln(M - P)\right].$$

Substituting into the previous expression and simplifying obtain

$$\ln\left[\left(\frac{x}{M - x}\right)\left(\frac{M - P}{P}\right)\right] = KMt$$

which can be solved to obtain the exact response expression

$$x(t) = \frac{M}{1 + ((M - P)/P)e^{-KMt}}.$$

The earlier observation that for any initial state, $x(t) \to M$ as $t \to \infty$ is validated by this analytical response.

For purpose of comparison an approximation to the exact time response is now determined using the linearization approach. Since $x = M$ is a constant equilibrium response define the variable z

$$z = x - M,$$

where z represents the variation or change of the resource level from the equilibrium level M. Thus

$$\frac{dz}{dt} = - KMz - Kz^2.$$

Now for x sufficiently near M, i.e., for z sufficiently small, the term Kz^2 in the above equation can be ignored in comparison with the term KMz; hence the linearized first order model is

$$\frac{dz}{dt} = - KMz$$

which can serve as an approximation to the original state model. If $x(0) = P$ then $z(0) = P - M$ so that, solving the above linearized equation, obtain

$$z(t) = (P - M)e^{-KMt}.$$

Thus, in terms of the resource level,

$$x(t) = M + (P - M)e^{-KMt}$$

which is an approximation to the exact response expression based on the above linearized model. Such a linearization approach is a valid way of determining approximate expressions for the time response, especially in cases where the exact response is difficult or impossible to obtain.

The significance of these particular results is worth comment. Without harvesting, the fisheries resource tends to the level M; this level is sometimes referred to as the carrying capacity of the environment. If the resource level is less than M the resource tends to increase asymptotically to the level M; if the resource level exceeds M the resource tends to decrease asymptotically to the level M. This phenomenon is sometimes referred to as a logistic growth effect and is characteristic of many renewable resources. This completes our analysis of the fisheries resource with no harvesting.

Constant Harvesting Rate. Now consider the case of a constant harvesting rate $u(t) = H$, $t \geqslant 0$. Thus the rate of change of the resource level is

$$\frac{dx}{dt} = Kx(M - x) - H.$$

It is convenient to plot the rate of change of the resource level dx/dt versus the resource level x as indicated in Figure 5-1-2.

As usual the fisheries resource is in equilibrium if $dx/dt = 0$ which holds if

$$Kx(M - x) = H.$$

There are two cases to be considered, as indicated in Figure 5-1-2.

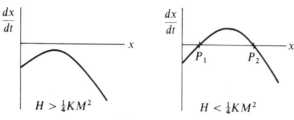

Figure 5-1-2

First suppose that $H > KM^2/4$ so that there is no intersection and hence no equilibrium state. In this case the harvesting rate always exceeds the renewal rate, $dx/dt < 0$, so that for any initial resource level the resource is eventually depleted to zero. Such a harvesting policy is clearly an unwise management policy.

Now consider a harvesting policy such that $0 < H < KM^2/4$; then as indicated in Figure 5-1-2 there are two intersections and hence two equilibrium states, denoted by P_1 and P_2, with $P_1 < P_2$. From the state equation it should be clear that $dx/dt > 0$ if $P_1 < x < P_2$, while $dx/dt < 0$ if $0 < x < P_1$ or $x > P_2$; in the former case the resource level is increasing, in the latter case it is decreasing. Thus, if the initial resource level $x(0) = P$ satisfies $0 < P < P_1$ then the resource is necessarily depleted to zero. If the initial resource level satisfies $P_1 < P$ then the resource level tends asymptotically to the constant level P_2. Obviously, it is this latter situation which is desirable from a management perspective. In fact, the most desirable management policy, based on a constant harvesting rate, is to maximize the harvesting rate H subject to satisfaction of the constraint that there is a maintainable resource, i.e., that $H \leq KM^2/4$. The maximum sustainable harvesting rate is clearly $H = KM^2/4$ and in this particular situation $P_1 = P_2 = (M/2)$. If the initial resource level $x(0) = P$ satisfies $P > M/2$ then the resource decreases asymptotically to the level $M/2$. However, if the initial resource level satisfies $0 < P < M/2$ then the resource is eventually depleted to zero. Thus this maximum sustainable harvesting rate $H = KM^2/4$ is appropriate as a management policy only for relatively large initial resource level satisfying $P > M/2$. For a smaller initial resource level where $P < M/2$ biologists say that the resource is overexploited since the resource would eventually be depleted if continued constant harvesting occurs. Explicit analytical solution responses are not determined in this case since the important qualitative features of the response are easily determined without the exact responses.

Constant Effort Harvesting. Finally, consider the situation referred to as a constant effort harvesting policy. If a constant effort is made then the actual harvesting rate is proportional to the level of the resource, i.e., with a constant effort the actual catch is proportional to the level of availability of the resource. Consequently, this constant effort harvesting policy corresponds to the harvesting rate $u(t) = Ex(t)$, $t \geq 0$, where E is a constant. In

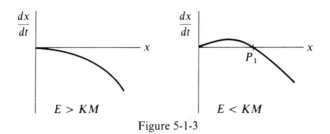

Figure 5-1-3

this case the rate of change of the resource is

$$\frac{dx}{dt} = Kx(M - x) - Ex.$$

In Figure 5-1-3 the rate of change of the resource level dx/dt is plotted versus the resource level x for the cases $E > KM$ and $E < KM$.

As usual the fisheries resource is in equilibrium when $dx/dt = 0$ which holds when

$$Kx(M - x) = Ex,$$

i.e., when the renewal rate of the resource equals the harvest rate.

As indicated in Figure 5-1-3 there are two cases to consider depending on the value of the effort constant E.

If there is a large harvesting effort so that $E > KM$ then the rate of change $dx/dt = 0$ only when $x = 0$; consequently in this case there is a single equilibrium state. Further, the harvesting rate always exceeds the renewal rate so that $dx/dt < 0$. Hence for any initial resource level the resource is eventually depleted to zero.

Now consider the smaller harvesting effort such that $E < KM$; then, as indicated in Figure 5-1-3, there are two equilibrium states, at $x = 0$ and $x = P_1$. Now $dx/dt > 0$ if $0 < x < P_1$ and $dx/dt < 0$ if $x > P_1$. Thus, for any initial nonzero resource level $x(t) \to P_1$, as $t \to \infty$, i.e., the resource level tends asymptotically to the level P_1. It is this harvesting policy, with $E < KM$, which is preferable in terms of maintaining a sustainable resource. The most desirable management policy, based on a constant effort harvesting policy, is to choose the effort constant E to maximize the harvesting rate EP_1, subject to satisfaction of the constraint that there should be a maintainable resource, i.e., that $E < KM$. This maximizing effort is easily calculated to be given by $E = KM/2$. The corresponding equilibrium resource level is $P_1 = M/2$ so that the resulting maximum harvesting rate, at equilibrium, is $EP_1 = KM^2/4$. Note that this maximum harvesting rate is the same as was obtained with the previously considered management policy of a constant harvesting rate. Notice, however, that in this case of constant effort harvesting with $E = KM/2$ that for any initial resource level the resource tends asymptotically to $M/2$, i.e., one-half of the carrying capacity. That is, overexploitation of the resource is not possible. This occurs due to the fact that if the resource level becomes small, with a constant fishing effort less of the resource will subsequently be harvested thus giving the resource a chance to renew itself. This is a good illustration of the advantage in using a feedback harvesting rate which adapts itself to the available resource level.

Case Study 5-2: Liquid Level in a Leaky Tank

Consider a leaky tank containing a liquid, as shown schematically in Figure 5-2-1.

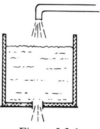

Figure 5-2-1

It is assumed that the tank has a hole in the bottom so that liquid flows out of the tank through the hole; liquid is also added into the tank as indicated; the flow rate Q into the tank is taken to be the input variable. The depth of water h in the tank is taken to be the output variable. All possible inertial effects of the liquid in the tank are ignored. Our interest is to determine how the flow rate affects the depth of liquid in the tank.

Letting h denote the depth of liquid in the tank it is easy to relate the velocity of liquid through the hole to the liquid depth. Based on a simple conservation of energy argument it is easy to show that the liquid velocity through the hole is

$$V = 0.6 \sqrt{2gh} ,$$

where g is the local acceleration of gravity constant and the dimensionless factor 0.6 is included to take into account friction effects of the hole. Now, the principle of conservation of mass can be used to characterize the rate of change of the depth of liquid in the tank. If the tank is assumed to have constant cross sectional area A_s and the area of the hole is A_h, then necessarily

$$A_s \frac{dh}{dt} = Q - A_h 0.6 \sqrt{2gh} ;$$

the expression on the left denotes the total rate of change of the volume of liquid in the tank; Q denotes the flow rate into the tank and the second term is the flow rate out of the tank through the hole. The variables Q and h are always assumed to be nonnegative to correspond to the physical requirements of the problem.

By choosing the depth of water as the state variable $x = h$ and the input variable u as the flow rate Q the state equation is

$$\frac{dx}{dt} = -0.6\frac{A_h}{A_s}\sqrt{2gx} + \frac{u}{A_s}.$$

This is clearly a first order nonlinear state model.

A few simple observations about this state model are first made. If there is no liquid in the tank initially, $x(0) = 0$, and there is no flow into the tank, $u(t) = 0$, $t \geqslant 0$, then clearly from the state model $x(t) = 0$, $t \geqslant 0$, i.e., there is never any liquid in the tank. Further if initially $x(0) \geqslant 0$ then $x(t) \geqslant 0$ for all $t \geqslant 0$. Thus the state model forces satisfaction of the required signs of the variables.

Now consider the important case where there is a constant flow rate $u(t) = \bar{Q}$, $t \geqslant 0$ into the standpipe; the state equation is thus

$$\frac{dx}{dt} = -0.6\frac{A_h}{A_s}\sqrt{2gx} + \frac{\bar{Q}}{A_s}.$$

This nonlinear differential equation can be integrated exactly by separation of variables; however the resulting analytical expressions are particularly messy and not very useful. Hence alternate analysis is pursued. First, it is easy to determine an equilibrium solution by requiring $dx/dt = 0$. Hence an equilibrium state corresponds to a solution of

$$-0.6\frac{A_h}{A_s}\sqrt{2gx} + \frac{\bar{Q}}{A_s} = 0,$$

so that

$$\bar{x} = 1.4\frac{\bar{Q}^2}{A_h^2 g}$$

is the unique equilibrium state. Thus if $x(0) = \bar{x}$ and $u(t) = \bar{Q}$, $t \geqslant 0$ then $x(t) = \bar{x}$ for all $t \geqslant 0$.

The above equilibrium state can be used to determine the solution responses if x is near $\bar{x}$ and u is near $\bar{Q}$ using the linearization procedure described earlier. Define the new variables

$$v = u - \bar{Q},$$

$$z = x - \bar{x} = x - 1.4\frac{\bar{Q}^2}{A_h^2 g}.$$

Substituting into the state equation obtain

$$\frac{dz}{dt} = -0.6\frac{A_h}{A_s}\sqrt{2g(z + \bar{x})} + \frac{1}{A_s}(v + \bar{Q}).$$

Now the nonlinear square root function can be approximated by a linear function in z near $z = 0$. Using the approach indicated in Appendix D obtain the approximation

$$\sqrt{2g(z + \bar{x})} \cong \sqrt{2g\bar{x}} + \sqrt{\frac{g}{2\bar{x}}}\, z.$$

Substituting into the previous expression and simplifying obtain

$$\frac{dz}{dt} = -0.3 \frac{A_h}{A_s} \sqrt{\frac{2g}{\bar{x}}}\, z + \frac{v}{A_s}$$

and using the expression for $\bar{x}$

$$\frac{dz}{dt} = -0.36 \left(\frac{A_h^2}{A_s} \frac{g}{\bar{Q}} \right) z + \frac{v}{A_s} .$$

As expected, the equation in the variation z is now a linear differential equation. It is important to keep in mind that this equation is an approximation to the previous state model which is valid as long as z is small compared to $\bar{x}$. As an example of the use of the linearized state model consider the determination of the response properties, assuming $x(0) \neq \bar{x}$ and $u(t) = \bar{Q}$, $t \geqslant 0$. The zero input response of the linearized model can be used to determine the corresponding response properties. Clearly, with $v(t) = 0$, $t \geqslant 0$ the linearized state equation has the solution

$$z(t) = e^{-t/\tau} z(0),$$

where $\tau = 2.8 A_s \bar{Q} / A_h^2 g$ so that

$$x(t) = \bar{x} + (x(0) - \bar{x}) e^{-t/\tau}.$$

Thus for any initial state $x(0)$ near $\bar{x}$, $x(t) \to \bar{x}$ as $t \to \infty$, i.e., the depth of liquid in the standpipe tends asymptotically to its equilibrium value. Thus the equilibrium state $\bar{x}$, corresponding to the constant input $u(t) = \bar{Q}$, $t \geqslant 0$, is said to be stable.

Case Study 5-3: High Temperature Oven

In this example a model is developed for the temperature inside an industrial high temperature oven. The effects of heat transfer to the oven via a heating element are considered and the effects of thermal losses from the oven to the surroundings are taken into account. The temperature T_0 of the oven is assumed to be uniform and the temperature T_s of the air surrounding the oven is assumed to be uniform. Let C denote the thermal capacity of the oven; let Q_i denote the heat transfer rate to the oven from the heater element and Q_0 denote the heat transfer rate due to the thermal losses from the oven; then

$$C \frac{dT_0}{dt} = Q_i - Q_0.$$

The thermal losses from the oven are due to natural convection and, since it is assumed that the operating temperature of the oven is high, it is also important to take into account thermal radiation losses. A reasonable expression for the total heat transfer losses, which can be validated via experimentation, is that

$$Q_0 = K_c(T_0 - T_s) + K_R(T_0^4 - T_s^4),$$

where the first term characterizes the convective losses and the second term characterizes the radiative losses; the constants K_c and K_R are the convective and radiative parameters. In the situation of interest the heat transfer rate Q_i is the input variable u; the output variable y is the temperature T_0 of the oven. The temperature of the oven can be chosen as the state variable so $x = T_0$. The first order nonlinear state model is given by

$$\frac{dx}{dt} = -\frac{K_c}{C}(x - T_s) - \frac{K_R}{C}(x^4 - T_s^4) + \frac{u}{C}.$$

It is important to recognize that the temperatures must necessarily be expressed in absolute temperature units.

It would be extremely difficult to determine the general response analytically as it depends on the initial state and the input function. Hence, some special situations are considered.

Consider the case where there is a constant heat transfer rate to the oven $u(t) = \overline{Q}$, $t > 0$. Then an equilibrium state, if there is one, necessarily satisfies the algebraic equation.

$$0 = -K_c(x - T_s) - K_R(x^4 - T_s^4) + \overline{Q}.$$

This nonlinear equation cannot be solved explicitly but it is possible to demonstrate the existence of a solution using graphical means. In Figure 5-3-1 the two functions $K_R(x^4 - T_s^4)$ and $\overline{Q} - K_c(x - T_s)$ are indicated as a function of the variable x; the indicated point where an intersection occurs corresponds to a solution of the above nonlinear algebraic solution and hence to an equilibrium state, denoted by $\overline{T}$.

Thus there is a single equilibrium state $\overline{T}$ with $\overline{T} > T_s$. Thus, if $u(t) = \overline{Q}$, $t \geqslant 0$ and $x(0) = \overline{T}$ then $x(t) = \overline{T}$, $t \geqslant 0$.

A linearized state model is now obtained which is valid so long as the temperature of the oven is not far different from the equilibrium temperature.

It is convenient to introduce the variables v and z defined by

$$u = \overline{Q} + v,$$
$$x = \overline{T} + z$$

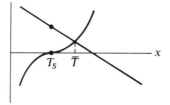

Figure 5-3-1

so that in terms of the variational variables v and z

$$\frac{dz}{dt} = \frac{-K_c}{C}(\overline{T} + z - T_s) - \frac{K_R}{C}\left[(\overline{T} + z)^4 - T_s^4\right] + \frac{\overline{Q} + v}{C} \; ;$$

or after expanding in a power series in z obtain

$$\frac{dz}{dt} = -\left[\frac{K_c}{C} + \frac{4K_R\overline{T}^3}{C}\right]z + \frac{v}{C} - \frac{K_c}{C}(\overline{T} - T_s)$$
$$- \frac{K_R}{C}(\overline{T}^4 - T_S^4) + \frac{\overline{Q}}{C} - \frac{K_R}{C}\left[8\overline{T}^2z^2 + 4\overline{T}z^3 + z^4\right].$$

Now by using the fact that $\overline{T}$ is an equilibrium state and by assuming that z is small so that the terms involving z^2, z^3, z^4 can be ignored the linearized state model

$$\frac{dz}{dt} = -\left[\frac{K_c}{C} + \frac{4K_R\overline{T}^3}{C}\right]z + \frac{v}{C}$$

is obtained. It would now be a relatively easy matter to use this linearized approximate model to ascertain the responses of the nonlinear system.

As one illustration, consider the case where $u(t) = \overline{Q}$, $t > 0$ and $x(0)$ is near $\overline{T}$. From the above linearized model it follows that $z(t) \to 0$ as $t \to \infty$ and consequently $x(t) \to \overline{T}$ as $t \to \infty$.

Note also that the rate at which $x(t) \to \overline{T}$ increases as $\overline{T}$ increases, i.e., the rate at which the oven temperature tends to its equilibrium value increases as that equilibrium value increases; this effect is completely due to the inclusion of the thermal radiation losses in the model of the oven.

The linear model could be used for other analyzes as well, so long as the assumption that z is small compared to $\overline{T}$ is valid. Note that since no terms involving the input were ignored in obtaining the linearized model it is not necessary to require that the input $u(t)$ is near $\overline{Q}$ in order to use the linearized model as an analytical tool.

Case Study 5-4: Circuit with Diode

In this example a simple circuit containing a capacitor and a diode element, i.e., a nonlinear resistor, is considered; a schematic picture of the circuit is indicated in Figure 5-4-1.

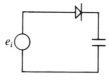

Figure 5-4-1

The diode has the property that its resistance to current flow depends on the voltage polarity across its terminals. Its current-voltage relationship is assumed to be given by

$$i_d = v_d / R_1 \quad \text{if } v_d \geqslant 0,$$
$$i_d = v_d / R_2 \quad \text{if } v_d < 0$$

where $0 < R_1 < R_2$, where i_d is the current through the diode and v_d is the voltage drop across the diode. This current-voltage characteristic is shown in Figure 5-4-2.

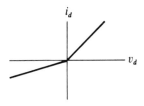

Figure 5-4-2

The capacitor is supposed to have a capacitance C. The input to the circuit is the voltage e_i; the output of the circuit is the voltage e_0 across the capacitance.

Clearly the current through the diode is the same as the current through the capacitor so that

$$C \frac{de_0}{dt} = i_i$$

and from Kirchoff's voltage law

$$e_0 + v_d - e_i = 0.$$

Consequently, it follows that

$$i_d = \frac{1}{R_1} (e_i - e_0) \quad \text{if } e_i - e_0 \geqslant 0,$$

$$i_d = \frac{1}{R_2} (e_i - e_0) \quad \text{if } e_i - e_0 < 0.$$

State equations are easily obtained by choosing the voltage across the capacitor as the state variable, $x = e_0$; the input variable u is the voltage e_i.

Thus the first order nonlinear state model

$$\frac{dx}{dt} = \frac{-x}{R_1 C} + \frac{u}{R_1 C} \quad \text{if } u - x \geqslant 0,$$

$$\frac{dx}{dt} = \frac{-x}{R_2 C} + \frac{u}{R_2 C} \quad \text{if } u - x < 0$$

is obtained. The response of the system to each of two different input functions is now determined.

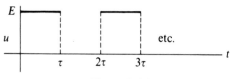

Figure 5-4-3

First, consider a constant input function $u(t) = E$, $t \geqslant 0$. If $x(0) = E$ then clearly $x(t) = E$, $t \geqslant 0$ so that $x = E$ is an equilibrium state. If $E - x(0) \geqslant 0$ then it is easily observed that $E - x(t) \geqslant 0$, $t \geqslant 0$ and that $x(t) \to E$ as $t \to \infty$. Similarly, if $E - x(0) \leqslant 0$ then $E - x(t) \leqslant 0$, $t \geqslant 0$ and $x(t) \to E$ as $t \to \infty$. Thus for any $x(0)$, $x(t) \to E$ as $t \to \infty$; but note that the decay rate depends on whether $x(0) < E$ or $x(0) > E$.

Now consider a periodic square wave input as shown in Figure 5-4-3. This square wave input has a period 2τ; it is desired to determine if there is some initial state for which the resulting response is periodic with the same period 2τ. The succeeding analysis demonstrates how to determine such a periodic response. If $x(0) < E$, then necessarily for $0 < t < \tau$,

$$\frac{dx}{dt} = \frac{-x}{R_1 C} + \frac{E}{R_1 C}$$

so that

$$x(t) = E - (E - x(0))\exp(-t/R_1 C) \quad \text{for } 0 \leqslant t \leqslant \tau$$

and

$$x(\tau) = E - (E - x(0))\exp(-\tau/R_1 C).$$

Thus for $\tau < t < 2\tau$, it follows that

$$\frac{dx}{dt} = \frac{-x}{R_2 C}$$

so that

$$x(t) = x(\tau)\exp(-(t - \tau)/R_2 C), \quad \tau \leqslant t \leqslant 2\tau,$$

or using the expression for $x(\tau)$

$$x(t) = \left[E - (E - x(0))\exp(-\tau/R_1 C) \right]\exp(-t/R_2 C), \quad \tau \leqslant t \leqslant 2\tau.$$

Now the response is periodic if

$$x(2\tau) = x(0),$$

that is, if

$$x(0) = \left[E - (E - x(0))\exp(-\tau/R_1 C) \right]\exp(-\tau/R_2 C).$$

Solving for $x(0)$ obtain

$$x(0) = \left[\frac{1 - \exp(-\tau/R_1 C)}{\exp(\tau/R_2 C) - \exp(-\tau/R_1 C)} \right] E.$$

Thus for this particular initial state it follows that the response of the system is periodic; a typical response is indicated as shown in Figure 5-4-4.

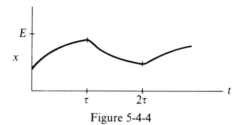

Figure 5-4-4

Note that the response is periodic but the waveform is rather complicated. It can be shown that for any initial state the response tends to the
above periodic waveform.

As indicated in this case study, it is sometimes possible for certain
"piecewise linear" nonlinear systems to determine the responses for particular input functions. In this first order example the difficulty was not too
great; in many other examples the details might be difficult to carry out.

A simple computer simulation has been performed for this case study
example; the program and results are given in Appendix E.

Exercises

5-1. Consider the fisheries model as discussed in Case Study 5-1.

   (a) Assuming a constant harvesting rate with

$$H < 0.25 KM^2$$

   develop a linearized model for the change in resource level. Use
   the linearized model to characterize the asymptotic response of
   the fisheries resource.

   (b) Assuming a constant effort harvesting policy with

$$E < KM$$

   develop a linearized model for the change in resource level. Use
   the linearized model to characterize the asymptotic response of
   the fisheries resource.

5-2. Consider the leaky tank model as described in Case Study 5-2 with
   the parameter values

$$A_s = 20.0 \text{ in}^2, \quad A_n = 1.0 \text{ in}^2.$$

   (a) If the tank is initially empty and there is a constant inflow rate
   $Q(t) = 30.0 \text{ in}^3/\text{sec}$ for $t \geqslant 0$ what is the liquid level in the tank
   thereafter for $t \geqslant 0$?

   (b) If initially the liquid level in the tank is 12.0 in and there is a
   constant inflow rate $Q(t) = 30.0 \text{ in}^3/\text{sec}$ for $t \geqslant 0$ what is the
   liquid level in the tank thereafter for $t \geqslant 0$?

(c) If initially the liquid level in the tank is 6.0 in and there is no flow into the tank what is the liquid level in the tank thereafter for $t \geqslant 0$? How long does it take for the tank to empty?

5-3. Consider the model of a high temperature oven as described in Case Study 5-3 with parameter values

$$T_s = 530°R,$$

$$C = 24.0 \text{ BTU}/°R,$$

$$K_c = 8.0 \text{ BTU}/(\text{hr-}°R),$$

$$K_R = 2.0 \times 10^{-8} \text{ BTU}/(\text{hr-}°R^4).$$

(a) What is the temperature in the oven for $t \geqslant 0$ if the initial temperature is 530°R and the heat transfer rate to the oven (from the heater) is 12,000 BTU/hr?

(b) What is the temperature in the oven for $t \geqslant 0$ if the initial temperature is 1000°R and there is no heat transfer to the oven, i.e., the heater is off?

5-4. Suppose that the parameter values for the diode circuit model developed in Case Study 5-4 are given by

$$R_1 = 5000.0 \text{ ohm}, \quad R_2 = 10000.0 \text{ ohm}, \quad C = 10^{-6} \text{ farad}.$$

(a) For a constant input voltage $e_i(t) = 200.0$ volts for $t \geqslant 0$, what is the circuit response if initially the capacitor voltage is 0.0 volts?

(b) For a constant input voltage $e_i(t) = 200.0$ volts for $t \geqslant 0$, what is the circuit response if initially the capacitor voltage is 300.0 volts?

(c) For a sinusoidal input voltage $e_i(t) = 200.0 \cos 1000t$ volts, $t \geqslant 0$, what is the steady state circuit response?

(d) For a periodic piecewise constant voltage input with

$$e_i(t) = 200.0 \text{ volts for } 0 \leqslant t < 0.01 \text{ sec},$$

$$e_i(t) = 0.0 \text{ volts for } 0.01 \leqslant t < 0.02 \text{ sec},$$

what is the steady state periodic circuit response?

5-5. A spherical drop of liquid is assumed to evaporate at a rate proportional to its surface area. Develop a state model for the radius of the drop. If initially the radius of the drop is 0.05 in and in 5.0 min its radius is 0.025 in what is the radius of the drop as a function of time? When does the drop completely evaporate?

5-6. The weight of a ball is 0.4 lb. There is air drag which retards the motion of the ball as it falls under a constant gravitational force. The drag force is assumed to be proportional to the square of the velocity of the ball; this force is 0.21 lb when the velocity of the ball is 2.0 ft/sec. Develop a state model for the velocity of the ball as it falls. If the ball is dropped from rest find its velocity as a function of time.

5-7. A vertical cylindrical vessel has a diameter of 1.8 meters and a height of 2.45 meters. The vessel has a hole in the bottom of diameter 6.0 centimeters. Assuming the vessel contains some amount of liquid which can flow out of the hole in the bottom of the vessel develop a state model for the depth of liquid in the vessel. If the vessel is initially full what is the depth of liquid in the vessel as a function of time?

5-8. A certain chemical dissolves in water at a rate proportional to the product of the amount undissolved and the difference between the concentration in a saturated solution and the concentration in the actual solution. The concentration is defined as a ratio of the mass of the chemical to the mass of the water. A saturated solution is assumed to consist of equal amounts of the chemical and water. Assume that initially 30.0 grams of the chemical are mixed with 100.0 grams of water and that 10.0 grams of the chemical are dissolved in 2.0 hr. Give a state model for the amount of dissolved chemical. How much of the chemical is dissolved as a function of time?

5-9. A compound C is formed by the reaction of two elements A and B in a closed container. One gram of A and one gram of B react to form two grams of C. Initially there are ten grams of A, five grams of B and none of compound C present in a reaction vessel. Assume that the rate of formation of compound C is proportional to the product of the masses of A and B uncombined. Develop a state model for the mass of chemical compound C formed by the reaction. Express the mass of compound C formed as a function of time. Assume that at the end of 2 min exactly one gram of compound C has been formed from the reaction.

5-10. A vertical cylindrical tank of volume V ft^3 and cross sectional area A ft^2 is initially full of water and is being drained by two pumps. Each pump operates at the constant flow rate of Q ft^3/min. One pump is located at the bottom of the tank; the other pump is located half of the way up the tank.

(a) Derive a state model for the depth of water in the tank.
(b) How long does it take for the tank to empty?
(c) What is the depth of water in the tank as a function of time?

5-11. An electromechanical transducer with a movable iron mass is indicated schematically below.

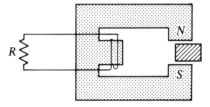

As the movable mass moves within the magnetic field developed by the permanent magnet a voltage is generated which causes a current flow in the circuit indicated. The circuit is assumed to be connected to a pure resistance. Let λ denote the magnetic flux in the air gap between the magnetic poles. Let z denote the position of the movable mass, measured from some reference position. Let i denote the current in the circuit. Then from Kirchoff's laws it follows that

$$0 = iR + \frac{d\lambda}{dt} \, .$$

The magnetic flux and current are related by the relation

$$\lambda = Li,$$

where the circuit resistance R is constant and the inductance L depends on the position of the movable mass according to

$$L = L_0\left(\frac{D}{D + z} \right).$$

Here L_0 and D are constants. Let the position of the movable mass be the input; let the voltage across the resistance R be the output.

(a) Give a state model for the transducer.
(b) If the input $z(t) = 0$ for $t \geqslant 0$ what are the equilibrium states for the transducer? If the input is any arbitrary constant value for $t \geqslant 0$ what are the equilibrium states for the transducer?
(c) What are linearized state equations which hold for small current flow in the circuit?
(d) Are the linearized equations in (c) stable? What is the physical interpretation of your conclusion?
(e) What is the steady state output of the transducer if the input motion is $z(t) = Z \cos \omega t$ for $t \geqslant 0$?

5-12. A simple capacitive circuit containing an ideal voltage source and a diode element is indicated below.

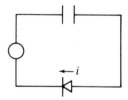

Let v denote the voltage supplied by the ideal voltage source; let C denote the capacitance in the circuit and let i denote the loop current. If v_d denotes the voltage drop across the diode then the diode characteristic is given by

$$i = I\big[\exp(\lambda v_d) - 1\big],$$

where I and λ are positive constants. Let the voltage v be the input variable; let the voltage drop v_d across the diode be the output variable.

(a) Write state equations for the circuit.
(b) If the input voltage $v(t) = 0$ for $t \geqslant 0$ what are the equilibrium states for the circuit?
(c) What are linearized state equations for the circuit in the case of small current flow in the circuit?
(d) Are the linearized equations in (c) stable? What is the physical interpretation of your conclusion?
(e) If the input voltage is $v(t) = V \sin \omega t$ for $t \geqslant 0$ what is the steady state output voltage of the circuit?

Chapter VI

Second Order Nonlinear State Models

In this chapter second order nonlinear state models are considered. As indicated in the previous chapter it is not possible to develop a general procedure which is always applicable in the analysis of such state models. Rather, a few special techniques, applicable to certain cases, can be presented. As in the previous chapter emphasis is given to simple integration techniques as well as to the approximation of the nonlinear state model by a linear state model. But for second order systems, as seen previously in Chapter III, the phase plane is sometimes useful for describing certain responses properties. As before, a brief description of some ideas and concepts is first given; then several physical processes are examined where some of the concepts prove helpful.

Some Systems Theory

Consider the second order nonlinear state model

$$\frac{dx_1}{dt} = f_1(x_1, x_2, u),$$

$$\frac{dx_2}{dt} = f_2(x_1, x_2, u),$$

$$y = g(x_1, x_2, u),$$

where, as before, the model is written in terms of first order differential equations. Here u is the input variable, y is the output variable and x_1, x_2 denote the two state variables. The state model is defined by the three functions $f_1(x_1, x_2, u)$, $f_2(x_1, x_2, u)$ and $g(x_1, x_2, u)$. Most attention in the following comments is given to the two differential equations since that is the part of the model which causes most difficulties.

In a few rare cases it is possible to obtain analytical solution expressions for the differential equations. One approach is to recognize that

$$\frac{dx_2}{dx_1} = \frac{dx_2/dt}{dx_1/dt} = \frac{f_2(x_1, x_2, u)}{f_1(x_1, x_2, u)},$$

which, in a few cases, can be integrated to obtain an expression of the form

$$x_2 = \Psi(x_1, u),$$

a so-called first integral. Then if the equation

$$\frac{dx_1}{dt} = f_1(x_1, \Psi(x_1, u), u)$$

can be integrated solutions $x_1(t)$, and thus $x_2(t)$, can be explicitly obtained. Unfortunately this procedure cannot often be carried out. As will be seen shortly, however, the first integral is of some interest in its own right.

One special class of solutions is of particular interest, namely the constant solutions of the state equations. Assuming a constant input function $u(t) = \bar{u}$, $t \geq 0$; if

$$0 = f_1(\bar{x}_1, \bar{x}_2, \bar{u}),$$

$$0 = f_2(\bar{x}_1, \bar{x}_2, \bar{u})$$

hold for some $\bar{x}_1, \bar{x}_2$ it follows that if $x_1(0) = \bar{x}_1$, $x_2(0) = \bar{x}_2$, and $u(t) = \bar{u}$, $t \geq 0$, then necessarily $x_1(t) = \bar{x}_1$, $x_2(t) = \bar{x}_2$ for all $t \geq 0$, i.e., $\bar{x}_1, \bar{x}_2$ is an equilibrium state corresponding to the constant input $\bar{u}$; the corresponding output function is necessarily constant and is given by $y(t) = g(\bar{x}_1, \bar{x}_2, \bar{u})$ for all $t \geq 0$. Thus the equilibrium states can be determined by solving two nonlinear algebraic equations in two unknowns.

It is often of interest to approximate a nonlinear state model, if this is possible. One approach is the following. Suppose that $\bar{x}_1, \bar{x}_2$ is an equilibrium state corresponding to the constant input function $\bar{u}$; then the new variables v, w, z_1, z_2 can be defined by

$$u = \bar{u} + v,$$
$$y = g(\bar{x}_1, \bar{x}_2, \bar{u}) + w,$$
$$x_1 = \bar{x}_1 + z_1,$$
$$x_2 = \bar{x}_2 + z_2.$$

The previous nonlinear state model can be rewritten in the equivalent form

$$\frac{dz_1}{dt} = f_1(\bar{x}_1 + z_1, \bar{x}_2 + z_2, \bar{u} + v),$$

$$\frac{dz_2}{dt} = f_2(\bar{x}_1 + z_1, \bar{x}_2 + z_2, \bar{u} + v),$$

$$w = g(\bar{x}_1 + z_1, \bar{x}_2 + z_2, \bar{u} + v) - g(\bar{x}_1, \bar{x}_2, \bar{u}),$$

where v and w are new input and output variables and z_1 and z_2 are new state variables. The above nonlinear functions can be approximated by linear functions in the variables z_1, z_2, v near $z_1 = 0, z_2 = 0, v = 0$ using the approach discussed in Appendix D; the result is that

$$\frac{dz_1}{dt} = \left(\frac{\partial f_1}{\partial x_1} \right) z_1 + \left(\frac{\partial f_1}{\partial x_2} \right) z_2 + \left(\frac{\partial f_1}{\partial u} \right) v,$$

$$\frac{dz_2}{dt} = \left(\frac{\partial f_2}{\partial x_1} \right) z_1 + \left(\frac{\partial f_2}{\partial x_2} \right) z_2 + \left(\frac{\partial f_2}{\partial u} \right) v,$$

$$w = \left(\frac{\partial g}{\partial x_1} \right) z_1 + \left(\frac{\partial g}{\partial x_2} \right) z_2 + \left(\frac{\partial g}{\partial u} \right) v,$$

which is a linear state model. All of the partial derivatives above are evaluated at $\bar{x}_1, \bar{x}_2, \bar{u}$. If the functions $u(t)$ and $x_1(t), x_2(t)$ are near the constant functions $\bar{u}$ and $\bar{x}_1, \bar{x}_2$, i.e., the variables v and z_1, z_2 are small, then the above linearized state model is a good approximation to the previous nonlinear model. Consequently, under the stated assumptions for which the linearized state equations are valid, the linearized state model can be analyzed, as indicated in Chapter III, to determine the response properties of the system.

When the state model involves only two state variables, as is the case under consideration, it may be useful to employ certain phase plane techniques. As was discussed in Chapter III the responses of the system can be represented by trajectories in the phase plane; each time response corresponds to a particular trajectory.

Suppose that the input function is constant, $u(t) = \bar{u}$, $t \geqslant 0$. Consequently, for each $x_1(0), x_2(0)$ there is a unique solution $x_1(t)$ and $x_2(t)$ defined for $t \geqslant 0$, which defines a corresponding trajectory in the phase plane. Our interest here is to indicate how to obtain sufficient information so that certain typical trajectories can be sketched, approximately, in the phase plane. Such a graphical approach is appropriate since the trajectories are particularly useful for giving a qualitative indication of the response properties of the system with $u(t) = \bar{u}$, $t \geqslant 0$. All equilibrium states should first be determined since they define points in the phase plane which are trajectories. Then, as just indicated, a linearized state model may be derived which is valid, as long as trajectories are near the equilibrium state; such linearized state equations may be used, as indicated in Chapter III, to determine the qualitative properties of the trajectories near each equilibrium state. More generally it is recognized that the slope of the trajectory through (x_1, x_2) in the phase plane is given by

$$\frac{dx_2}{dx_1} = \frac{dx_2/dt}{dx_1/dt} = \frac{f_2(x_1, x_2, \bar{u})}{f_1(x_1, x_2, \bar{u})} .$$

In a few cases this first order differential equation can be integrated, e.g., by separation of variables, to determine an analytical expression for the trajectories. But in most cases the above relation is most helpful as an expression for the slope of the trajectories. For example, it is directly observed that the slope of a trajectory satisfies

$$\frac{dx_2}{dx_1} = 0 \quad \text{if} \quad f_2(x_1, x_2, \bar{u}) = 0,$$

$$\frac{dx_2}{dx_1} = \infty \quad \text{if} \quad f_1(x_1, x_2, \bar{u}) = 0;$$

thus the algebraic equations $f_1(x_1, x_2, \bar{u}) = 0$ and $f_2(x_1, x_2, \bar{u}) = 0$ each define the isocline curves along which the trajectories are horizontal and vertical respectively. Note that in general the isocline curves for nonlinear systems are not straight lines in the phase plane; they are more general curved lines. Other isocline curves can also be determined based on the above expression for the slope of the trajectories. Usually, sufficient information can be obtained using the above approaches so that a number of important typical trajectories can be sketched in the phase plane with the correct qualitative properties. A number of physical examples are considered subsequently which illustrate many of these ideas.

Case Study 6-1: Motion of a Boat Crossing a River

In this example the two dimensional motion of a boat crossing a river is considered. Assume a fixed coordinate system so that the location of the boat is specified by the cross stream and down stream coordinate values x_1 and x_2 as shown in Figure 6-1-1. The banks of the river, as indicated, correspond to the straight lines $x_1 = 0$ and $x_1 = L$. It is assumed that the boat moves with a constant velocity V relative to the water. As indicated in Figure 6-1-1 the relative velocity vector makes a heading angle α with the x_1 axis. There is a current in the river with velocity solely along the x_2 axis given by $W \sin(\pi x_1 / L)$. Note that the river current is zero at each bank where $x_1 = 0$ and $x_1 = L$, and the river current has the maximum velocity W at midstream where $x_1 = L/2$. In this example the heading angle α is the input variable and the coordinate location of the boat is the output variable of interest.

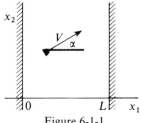

Figure 6-1-1

The equations describing the motion of the boat are obtained by using the fact that the absolute velocity of the boat is the velocity of the river current plus the velocity of the boat relative to the river. In each of the coordinate directions the velocity expressions are

$$\frac{dx_1}{dt} = V \cos \alpha,$$

$$\frac{dx_2}{dt} = V \sin \alpha + W \sin\left(\frac{\pi x_1}{L}\right).$$

This is a second order nonlinear state model for the motion of the boat in the river. The variables x_1 and x_2 are the state variables; the heading angle α is the input variable. This is a very simple model since the first state equation is uncoupled from the second state equation; thus the response for x_1 can be determined independently of the variable x_2.

The special case of a constant heading angle

$$\alpha(t) = \bar{\alpha}, \quad t \geqslant 0,$$

is now examined. As well, suppose that the coordinate axes are chosen, as indicated in Figure 6-1-1, so that the initial location of the boat is given by

$$x_1(0) = 0, \quad x_2(0) = 0.$$

The first state equation is easily solved to obtain

$$x_1(t) = [V \cos \bar{\alpha}] t.$$

Then the second state equation can be written as

$$\frac{dx_2}{dt} = V \sin \bar{\alpha} + W \sin\left(\frac{\pi V \cos \bar{\alpha} t}{L}\right).$$

This differential equation is easily solved to obtain

$$x_2(t) = [V \sin \bar{\alpha}] t + \frac{WL}{\pi V \cos \bar{\alpha}}\left[1 - \cos\left(\frac{\pi V \cos \bar{\alpha} t}{L}\right)\right].$$

The particular phase plane trajectory, which in this case is the path of the boat, is easily seen to be given by the analytical expression

$$x_2 = x_1 \tan \bar{\alpha} + \frac{WL}{\pi V \cos \bar{\alpha}}\left[1 - \cos\left(\frac{\pi x_1}{L}\right)\right].$$

This trajectory is shown in Figure 6-1-2. Thus the boat reaches the opposite bank of the river, $x_1(T) = L$, when

$$T = \frac{L}{V \cos \bar{\alpha}},$$

and the downstream location of the boat is given by

$$x_2(T) = L \tan \bar{\alpha} + \frac{2WL}{\pi V \cos \bar{\alpha}}.$$

In order for the boat to reach the opposite bank of the river with no change in its x_2 coordinate then the constant heading angle $\bar{\alpha}$ must be chosen so

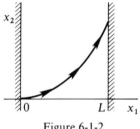

Figure 6-1-2

that $x_2(T) = 0$, i.e.,

$$\sin \bar{\alpha} = - \frac{2W}{\pi V} .$$

Clearly the constant heading angle must be negative. Further the condition that $x_2(T) = 0$ can only be achieved if $2W < \pi V$. Otherwise the river current is too strong.

Obviously, the model could be used to determine the responses for other than a constant heading angle. The basic procedure indicated would still be valid.

Case Study 6-2: Spread of an Epidemic

One important application of the system view point in the area of biology and medicine has been to the mathematical theory of epidemics. Consider the two classes of individuals in a fixed population: individuals who are susceptible to a certain disease and individuals who have the disease and hence are infectious. The disease is spread by direct contact between the susceptible individuals, or susceptibles, and the infective individuals, or infectives. It is assumed that an infective individual eventually becomes immune to the disease or dies from the disease; in any event an infective individual eventually loses his infectiousness.

Let $x_1(t)$ denote the number of susceptible individuals at time t and let $x_2(t)$ denote the number of infective individuals at time t. Each group is assumed to be homogeneous in that differences between individuals are ignored. The values for x_1 and x_2 can be any nonnegative number; for simplicity the values need not be integers; thus it may be suitable to think of numerical values of x_1 and x_2 as a spatial average or density. Descriptive equations for the epidemic are now obtained. If the contact rate between susceptibles and infectives is proportional to the product of the numbers of susceptibles and infectives then it is reasonable to assume that the rate at which susceptibles catch the disease and become infective is given by $\beta x_1 x_2$ where β is an infection parameter. In addition to the occurence of new cases of the disease the fact that infectives die or become immune from the disease is included; it is assumed that this rate is proportional to the number of infectives, i.e., to γx_2 where γ is a parameter related to the average time constant of the disease. As a final factor to be considered it is assumed that individuals may be removed from the susceptible class at a rate u which depends on vaccination efforts. Consequently, the equations which describe the dynamics of the epidemic are

$$\frac{dx_1}{dt} = -\beta x_1 x_2 - u,$$

$$\frac{dx_2}{dt} = \beta x_1 x_2 - \gamma x_2.$$

Clearly, the variables x_1 and x_2 constitute the state variables; the rate u is considered to be the input. Note that the epidemic model depends on the two parameters β and γ; and the state model is a second order nonlinear model.

First, consider the case where there is no vaccination effort so that $u(t) = 0$, $t \geqslant 0$. Consequently,

$$\frac{dx_1}{dt} = -\beta x_1 x_2,$$

$$\frac{dx_2}{dt} = \beta x_2 \left(x_1 - \frac{\gamma}{\beta} \right).$$

Any state x_1, x_2 with $x_2 = 0$ is thus an equilibrium state. Our first effort is to obtain sufficient information to sketch some typical trajectories in the phase plane, so that the qualitative features of the responses are clear. Thus the slope of a trajectory through x_1, x_2 is

$$\frac{dx_2}{dx_1} = \frac{\beta x_2(x_1 - \gamma/\beta)}{-\beta x_1 x_2} = \frac{(x_1 - \gamma/\beta)}{-x_1}.$$

This information can be used to graphically determine some of the trajectories; this approach is considered shortly but in this case it is actually possible to integrate the above equation to determine an analytical expression for the trajectories. Integrating via the separation of variables approach obtain

$$x_2 = -x_1 + \frac{\gamma}{\beta} \ln(x_1) + C,$$

where C is a constant of integration. Determining the constant C in terms of the initial state values $x_1(0) = N_S, x_2(0) = N_I$ obtain

$$x_2 - N_I = \frac{\gamma}{\beta} \ln\left(\frac{x_1}{N_S} \right) - (x_1 - N_S),$$

an analytical expression for the trajectories. Unfortunately this expression is rather complicated and does not directly indicate the qualitative features of the trajectories. Thus the previous expression for the slope of the trajectories is reconsidered.

First note that if $x_1 - (\gamma/\beta) = 0$ then $dx_2/dx_1 = 0$, i.e., the trajectories

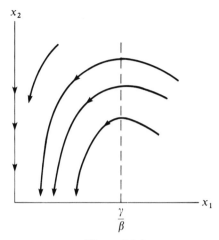

Figure 6-2-1

have a horizontal tangent. Next note that if $x_1 = 0$ then $dx_2/dx_1 = \infty$, i.e., the trajectories have a vertical tangent. Furthermore, by examination of the state equations it can be seen that if $x_1 \geq 0, x_2 \geq 0$ then $dx_1/dt \leq 0$; if $x_2 \geq 0$ and $0 \leq x_1 \leq \gamma/\beta$ then $dx_2/dt \leq 0$; if $x_2 \geq 0$ and $x_1 \geq \gamma/\beta$ then $dx_2/dt \geq 0$. Thus in the first two instances x_1 and x_2 decrease as time increases; in the third instance x_2 increases as time increases. Since our only interest, for physical reasons, is the region $x_1 \geq 0, x_2 \geq 0$ of the phase plane the above facts are sufficient for us to be able to sketch some typical trajectories in the phase plane. Several typical trajectories are indicated in Figure 6-2-1.

Several conclusions can be drawn from the trajectories indicated in Figure 6-2-1. First, note that in general $x_2(t) \rightarrow 0$ and $x_1(t)$ tends to some constant as $t \rightarrow \infty$, i.e., the number of infectives tends to zero but the number of susceptibles tends to some constant positive value less than γ/β. In addition if $x_1(0) < \gamma/\beta$ then the number of infectives decreases monotonically; however, if $x_1(0) > \gamma/\beta$ then the number of infectives initially increases during some time period and then decreases. The epidemic is at its peak, i.e., the number of infectives is at its maximum level, when $x_1 = \gamma/\beta$. These are only a few brief observations. But it is important to recognize that some important qualitative response properties can be determined without explicitly solving the state equations.

Now consider the response of the system if the input is not necessarily zero. It is not possible to indicate, generally, how the responses depend on the input but it is possible to look at a rather special case, namely where the number of susceptibles is near the value γ/β and the number of infectives is near zero. Recall that $x_1 = \gamma/\beta$ and $x_2 = 0$ is an equilibrium state corresponding to $u(t) = 0$, $t \geq 0$. The approach is to obtain a linear state model, which is valid under the stated assumptions, namely that x_1 is near γ/β and x_2 is near zero. In order to obtain such an approximation introduce new variables z_1 and z_2 defined by

$$z_1 = x_1 - \frac{\gamma}{\beta}, \quad z_2 = x_2 + 0.$$

Consequently

$$\frac{dz_1}{dt} = -\beta\left(z_1 + \frac{\gamma}{\beta}\right)z_2 - u,$$

$$\frac{dz_2}{dt} = \beta\left(z_1 + \frac{\gamma}{\beta}\right)z_2 - \gamma z_2.$$

Now assuming that z_1 and z_2 are small, i.e., x_1 is near γ/β and x_2 is near zero, the above nonlinear equations can be approximated by the linear state equations

$$\frac{dz_1}{dt} = -\gamma z_2 - u,$$

$$\frac{dz_2}{dt} = 0.$$

Now these relatively simple state equations can be used to determine the dependence of the responses on the input. In particular the general solution is easily seen to be

$$z_1(t) = z_1(0) - z_2(0)\gamma t - \int_0^t u(\tau)\,d\tau,$$

$$z_2(t) = z_2(0),$$

so that consequently if $x_1(0) = N_S$ and $x_2(0) = N_I$ then

$$x_1(t) = \left(N_S - \frac{\gamma}{\beta}\right) - N_I\gamma t - \int_0^t u(\tau)\,d\tau,$$

$$x_2(t) = N_I.$$

Obviously this approximation is not exact since it indicates that the number of infectives is not affected by vaccination effort; but the dependence of the number of susceptibles on the vaccination effort is in agreement with what is expected.

These brief comments about a particular model of an epidemic should demonstrate that the system viewpoint is applicable to a wide class of engineering and scientific problem solving efforts.

Case Study 6-3: Tunnel-Diode Circuit

An interesting and practical circuit containing a nonlinear resistive device, a tunnel diode, is shown in Figure 6-3-1. The current-voltage characteristic of the tunnel-diode element is also shown in Figure 6-3-1. One of the

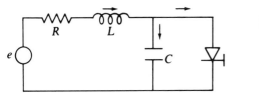 

Figure 6-3-1

practical uses of this circuit is as a computer circuit. Assume that the external voltage e is the input and the voltage drop across the tunnel diode v_d is the output; here $i_d = f(v_d)$ is the current through the tunnel diode element.

The Kirchoff voltage and current laws lead to the equations

$$i = i_d + C \frac{dv}{dt}$$

and

$$-e + Ri + L \frac{di}{dt} + v = 0,$$

where i is the current through the inductor and v is the voltage drop across the capacitor. In addition,

$$v_d - v = 0.$$

Using this result in the above equations obtain

$$\frac{dv}{dt} = -\frac{1}{C} f(v) + \frac{i}{C},$$

$$\frac{di}{dt} = -\frac{v}{L} - \frac{R}{L} i + \frac{e}{L}.$$

The voltage v and current i can be chosen as state variables, so that $x_1 = v$, $x_2 = i$; the input variable u is the applied voltage e. Thus the state variable equations are

$$\frac{dx_1}{dt} = -\frac{1}{C} f(x_1) + \left(\frac{1}{C} \right) x_2,$$

$$\frac{dx_2}{dt} = -\left(\frac{1}{L} \right) x_1 - \left(\frac{R}{L} \right) x_2 + \left(\frac{1}{L} \right) u,$$

where the function f is as indicated in Figure 6-3-1.

The case of most interest is where there is a constant input voltage $u(t) = E$, $t \geqslant 0$, where $E > 0$. The circuit is in equilibrium if the voltage x_1 and current x_2 satisfy

$$0 = -\left(\frac{1}{C} \right) f(x_1) + \left(\frac{1}{C} \right) x_2,$$

$$0 = -\left(\frac{1}{L} \right) x_1 - \left(\frac{R}{L} \right) x_2 + \frac{E}{L}.$$

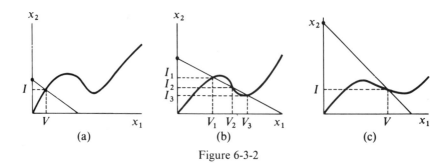

Figure 6-3-2

These simultaneous equations are easily reduced to satisfaction of the single equation

$$f(x_1) = \left(\frac{1}{R}\right)(E - x_1).$$

This single nonlinear algebraic equation characterizes the equilibrium voltage x_1 for the circuit. Although it is not possible to obtain an analytical expression for this equilibrium voltage unless an analytical expression for the function f is given, it is possible to indicate qualitatively what can happen. By indicating the solution of the above equilibrium equation graphically it is seen in Figure 6-3-2 that there are three distinct possibilities, depending on the particular values of the parameters. The three possibilities are characterized by the intersections shown in (a), (b), (c) of Figure 6-3-2. In case (a) there is one intersection, i.e., one equilibrium state V, I, and the slope of f at the intersection is positive. In case (b) there are three intersections, i.e., three equilibrium states $V_1, I_1; V_2, I_2; V_3, I_3;$ the slopes of the function f at the two extreme intersections are positive and at the intermediate intersection the slope is negative. In case (c) there is a single intersection, i.e., one equilibrium state V, I such that the slope of f at the intersection exceeds $-(1/R)$. Note that each intersection determines the equilibrium voltage V, which is easily determined graphically; the corresponding equilibrium current is of course $I = f(V)$. In all cases, the equilibrium voltage $V < E$ and the equilibrium current $I < E/R$. Each of the three cases, corresponding to (a), (b), (c) in Figure 6-3-2, is now examined. In each case the responses of the circuit are determined via an approximation of the nonlinear state equations near an equilibrium state by linearized state equations.

Consider the case (a) where V and I denote the unique equilibrium state. If the variable change

$$x_1 = V + z_1, \quad x_2 = I + z_2$$

is made, and $f(x_1)$ is approximated by the linear relation

$$f(x_1) \cong f(V) + K(x_1 - V),$$

where $K = df(V)/dx > 0$, then the resulting linearized state equations are

$$\frac{dz_1}{dt} = -\left(\frac{K}{C}\right)z_1 + \left(\frac{1}{C}\right)z_2,$$

$$\frac{dz_2}{dt} = -\left(\frac{1}{L}\right)z_1 - \left(\frac{R}{L}\right)z_2;$$

these state equations are valid so long as x_1 is near V. The characteristic polynomial for these linearized equations is

$$d(s) = s^2 + \left(\frac{R}{L} + \frac{K}{C}\right)s + \frac{1}{LC}(1 + RK).$$

The characteristic zeros always have negative real parts since $K > 0$, so that for $z_1(0)$ and $z_2(0)$ sufficiently small $z_1(t) \to 0$ and $z_2(t) \to 0$ as $t \to \infty$. In other words, $x_1(t) \to V$ and $x_2(t) \to I$ as $t \to \infty$. Thus, if case (a) holds the voltage and current tend asymptotically to their equilibrium values. Although it does not follow directly from the previous analysis, this result can be shown to hold for any inititial state. The equilibrium state V, I is thus said to be stable; the circuit, assuming case (a) holds, is called a *monostable circuit*.

Now consider case (b), as shown in Figure 6-3-1, where there are three distinct equilibrium states which are denoted by $V_1, I_1; V_2, I_2; V_3, I_3$ where $V_1 < V_2 < V_3$. Note that

$$K_1 = \frac{df}{dx}(V_1) > 0, \quad K_2 = \frac{df}{dx}(V_2) < 0, \quad K_3 = \frac{df}{dx}(V_3) > 0.$$

Linearized state equations are now obtained by approximation of the nonlinear state equations near each of the equilibrium states. Depending on which equilibrium state is considered, it can be shown that the linearized state equations about the equilibrium state V_i, I_i are

$$\frac{dz_1}{dt} = -\left(\frac{K_i}{C}\right)z_1 + \left(\frac{1}{C}\right)z_2,$$

$$\frac{dz_2}{dt} = -\left(\frac{1}{L}\right)z_1 - \left(\frac{R}{L}\right)z_2$$

for each $i = 1, 2, 3$; the characteristic polynomial is

$$d_i(s) = s^2 + \left(\frac{R}{L} + \frac{K_i}{C}\right)S + \frac{1}{LC}(1 + RK_i).$$

Now since $K_1 > 0$ and $K_3 > 0$ it follows that the equilibrium states V_1, I_1 and V_3, I_3 are stable, that is if $x_1(0)$ is near V_1 and $x_2(0)$ is near I_1 then $x_1(t) \to V_1$ and $x_2(t) \to I_1$ as $t \to \infty$, and if $x_1(0)$ is near V_3 and $x_2(0)$ is near I_3 then $x_1(t) \to V_3$ and $x_2(t) \to I_3$ as $t \to \infty$. Considering the characteristic polynomial for the linearized equations near V_2, I_2, it is clear from case (b) in Figure 6-3-2 that $K_2 < -(1/R)$ so that $1 + K_2 R < 0$; thus, in this case, there is a positive and a negative characteristic zero. Thus the equilibrium

Figure 6-3-3

state V_2, I_2 is said to be unstable. It can also be shown that for any initial state the voltage and current tend asymptotically to one of the stable equilibrium states. Since there are two stable equilibrium states the circuit is said to be a *bistable circuit*. Bistable circuits of this type are important computer circuits since a voltage pulse may cause a change from one stable equilibrium state to the other stable equilibrium state. Without going into details some typical trajectories in the phase plane are shown in Figure 6-3-3.

Finally, consider case (c) as shown in Figure 6-3-2. There is a single equilibrium state denoted by V, I. As before, the linearized state equations in the variations z_1, z_2 of the state from the equilibrium state V, I are

$$\frac{dz_1}{dt} = -\left(\frac{K}{C}\right)z_1 + \left(\frac{1}{C}\right)z_2,$$

$$\frac{dz_2}{dt} = -\left(\frac{1}{L}\right)z_1 - \left(\frac{R}{L}\right)z_2,$$

where $K = df(V)/dx$. The characteristic polynomial is

$$d(s) = s^2 + \left(\frac{R}{L} + \frac{K}{C}\right)s + \frac{1}{LC}(1 + RK).$$

Thus, as seen from Figure 6-3-2(c), $-RC/L < K$, so that $R/L + K/C > 0$. Hence the equilibrium state V, I is stable. The circuit is again said to be a monostable circuit.

Case Study 6-4: Underwater Launch of a Rocket

In this case study the vertical motion of a rocket fired from an underwater submarine is studied. The main feature of interest in the analysis is the variation in the resistance force depending on whether the rocket is in the water or in the air; the resistance force while the rocket is in the water is assumed to be proportional to the velocity of the rocket, while the resistance force while the rocket is in the air is ignored. In addition the variation of gravity with altitude is ignored. Only vertical motion of the rocket is considered. Let h denote the altitude of the rocket as measured from the surface of the water; if the rocket is in the water then $h < 0$ and if the rocket is in the air then $h > 0$. The three forces on the rocket are the weight W of the rocket, the resistance force, and a thrust or propulsive force T. From Newton's law the vertical motion of the rocket is described by

$$\frac{W}{g}\frac{d^2h}{dt^2} = T - \mu\frac{dh}{dt} - W \qquad \text{if} \quad h \leqslant 0,$$

$$\frac{W}{g}\frac{d^2h}{dt^2} = T - W \qquad \text{if} \quad h > 0.$$

The altitude and velocity of the rocket can be chosen as the state variables so that $x_1 = h$, $x_2 = dh/dt$; the input variable u is assumed to be the rocket thrust T. Thus the state model is

$$\frac{dx_1}{dt} = x_2,$$

$$\frac{dx_2}{dt} = \begin{cases} -\dfrac{\mu g}{W}x_2 + \left(\dfrac{g}{W}\right)u - g & \text{if} \quad x_1 \leqslant 0, \\[2mm] \left(\dfrac{g}{W}\right)u - g & \text{if} \quad x_1 > 0. \end{cases}$$

These are two nonlinear state equations; in this case dx_2/dt is in fact a discontinuous function of x_1 as indicated, where the discontinuity is at $x_1 = 0$.

In order to demonstrate how such a state model can often be analyzed, consider the special case for which the rocket is launched with $x_1(0) = -H < 0$ and with initial velocity $x_2(0) = V > 0$; Also assume there is no thrust so that $u(t) = 0$, $t \geqslant 0$. Other cases could be considered as well. The zero input response is now determined analytically.

Note that since $x_1(0) < 0$ there is some initial time period during which $x_1(t) \leqslant 0$; during that time period, which is denoted by $0 \leqslant t \leqslant t_1$,

$$\frac{dx_2}{dt} = \frac{-\mu g}{W}x_2 - g.$$

The solution during this time period is easily determined analytically; taking the Laplace transform of the state equations obtain

$$sX_1 + H = X_2,$$

$$sX_2 - V = \left(\frac{-\mu g}{W}\right)X_2 - \frac{g}{s}.$$

Consequently

$$X_1 = \frac{-H}{s} + \frac{V}{s\left(s + \dfrac{\mu g}{W}\right)} - \frac{g}{s^2\left(s + \dfrac{\mu g}{W}\right)}.$$

Taking the inverse transform obtain

$$x_1(t) = -H + \frac{W}{\mu g}\left\{\left(V + \frac{W}{\mu}\right)\left(1 - \exp\left(\frac{-\mu g t}{W}\right)\right) - gt\right\}$$

which is valid for $0 \leqslant t \leqslant t_1$, as long as the rocket is in the water. Now the rocket reaches the surface of the water when $x_1(t_1) = 0$, i.e.,

$$0 = -H + \frac{W}{\mu g} \left\{ \left(V + \frac{W}{\mu} \right) \left(1 - \exp\left(\frac{-\mu g t_1}{W} \right) \right) - g t_1 \right\}.$$

This is a transcendental equation for t_1 which cannot be solved analytically. But if t_1 were determined using numerical procedures then the velocity of the rocket at the surface of the water is

$$V_1 = x_2(t_1) = \left(V + \frac{W}{\mu} \right) \exp\left(-\frac{\mu g t_1}{W} \right) - \frac{W}{\mu}.$$

It is important to assume that $V_1 > 0$ holds since this guarantees that the rocket actually reaches the surface of the water.

Now since $x_1(t_1) = 0$ and $x_2(t_1) > 0$ it follows that the rocket is in the air for some time period after t_1, say $t_1 \leqslant t \leqslant t_2$. During this time period

$$\frac{dx_2}{dt} = -g.$$

Intergrating this equation and then the equation for dx_1/dt, and using the fact that $x_1(t_1)$ and $x_2(t_1)$ are given as before,

$$x_1(t) = V_1(t - t_1) - \tfrac{1}{2} g(t - t_1)^2$$

which is valid during the time period $t_1 \leqslant t \leqslant t_2$ while the rocket is in the air.

Note that the time t_2 when the rocket falls back into the water satisfies $x_1(t_2) = 0$, i.e.,

$$t_2 = t_1 + 2 \frac{V_1}{g}.$$

The altitude of the rocket as a function of time is indicated qualitatively in Figure 6-4-1. A simple computation shows that the rocket reaches a maximum altutude of $0.5(V_1^2/g)$ at time $t_1 + V_1/g$.

A simple computer simulation for this case study is given in Appendix E.

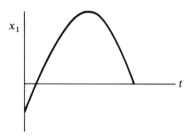

Figure 6-4-1

Case Study 6-5: A Study of Blood Sugar and Insulin Levels with Application to Diabetes

An important medical problem is that of the proper control of the blood sugar level in a diabetic person. One method for understanding mechanisms by which proper control can be achieved is through the development and study of a mathematical model for the relationship between the blood sugar level and the blood insulin level.

Let x_1 and x_2 be measures of the blood sugar level and blood insulin level in a given individual. Let u_1 and u_2 be measures of the food input rate and the insulin input rate for that individual. For example, x_1 and x_2 might be expressed in terms of the mass of blood sugar or blood insulin per unit volume of blood; u_1 and u_2 might be expressed in terms of mass per unit time. The objective is to determine how the two input variables u_1 and u_2 affect the two output variables x_1 and x_2; the relationship is clearly dynamic so that it can be described in terms of differential equations.

The biochemistry of an individual is that during a long fasting period, with $u_1 = 0, u_2 = 0$, there is an equilibrium with $x_1 = M_1$ and $x_2 = 0$, i.e., in equilibrium there is no insulin in the blood. Typically, the value for M_1 can be determined by a medical experiment. Expressions are now obtained for the rate of change of the blood insulin level and of the blood sugar level.

If the blood sugar level exceeds the equilibrium value M_1, $x_1 > M_1$ then insulin is secreted by the pancreas at a rate proportional to the difference $x_1 - M_1$; if $M_1 < x_1$ then no insulin is secreted by the pancreas. In addition, insulin in the blood undergoes a biochemical reaction and is biochemically reduced at a rate proportional to the insulin level x_2. Finally, the blood insulin level may increase, at least in the case of a diabetic, due to external addition of insulin directly to the blood, at a rate proportional to u_2. Thus the total rate of change of the blood insulin level is given by

$$\frac{dx_2}{dt} = a_3(x_1 - M_1) - a_4 x_2 + b_2 u_2 \qquad \text{if} \quad x_1 \geqslant M_1,$$

$$\frac{dx_2}{dt} = -a_4 x_2 + b_2 u_2 \qquad \text{if} \quad x_1 < M_1,$$

where a_3, a_4, b_2 are coefficients characteristic of the given individual's insulin utilization characteristics.

Now consider the rate of change of blood sugar level. The presence of insulin in the blood induces metabolism of the blood sugar thus reducing the blood sugar level; this rate of change of the blood sugar level is proportional to both the insulin and blood sugar levels. In this case the metabolism rate is assumed to be proportional to the product $x_1 x_2$. If the blood sugar level is smaller than its equilibrium level, $x_1 < M_1$, then sugar is released from the liver at a rate proportional to that difference $M_1 - x_1$; if $x_1 > M_1$ then no sugar is released by the liver. Finally, the blood sugar level may increase due to addition of sugar directly to the blood from eating, at a rate proportional to u_1. Consequently the total rate of change of the blood sugar level is given by

$$\frac{dx_1}{dt} = -a_1 x_1 x_2 + a_2(M_1 - x_2) + b_1 u_1 \qquad \text{if } x_1 \leqslant M_1,$$

$$\frac{dx_1}{dt} = -a_1 x_1 x_2 + b_1 u_1 \qquad \text{if } x_1 > M_1,$$

where a_1, a_2, b_1 are coefficients characteristic of the given individual's sugar utilization characteristics.

Thus, in summary, the second order nonlinear state model

$$\left. \begin{aligned} \frac{dx_1}{dt} &= -a_1 x_1 x_2 - a_2(x_1 - M_1) + b_1 u_1, \\ \frac{dx_2}{dt} &= -a_4 x_2 + b_2 u_2, \end{aligned} \right\} \quad \text{if } x_1 \leqslant M_1,$$

$$\left. \begin{aligned} \frac{dx_1}{dt} &= -a_1 x_1 x_2 + b_1 u_1, \\ \frac{dx_2}{dt} &= a_3(x_1 - M_1) - a_4 x_2 + b_2 u_2, \end{aligned} \right\} \quad \text{if } x_1 > M_1,$$

is obtained. Note that there are two nonlinear relationships in this model, one depending on the sign of $x_1 - M_1$ and the other involving the product $x_1 x_2$. Note also that the state equations are coupled. Consequently a complete mathematical analysis of the model is not possible. In what follows some of the major qualitative features of the responses are discussed; clearly a simulation study is required to add the details.

First, notice that if $u_1(t) = 0$, and $u_2(t) = 0$, $t \geqslant 0$, then $x_1(t) = M_1$ and $x_2(t) = 0$, $t \geqslant 0$ is an equilibrium response, consistent with the stated assumptions.

Now consider the particular case where there is only a food input which is described by the exponential form

$$u_1(t) = R_1 e^{-k_1 t}, \quad u_2(t) = 0, \qquad t \geqslant 0,$$

and the initial state is given by the equilibrium values

$$x_1(0) = M_1, \quad x_2(0) = 0.$$

Now $dx_1/dt = R_1 > 0$ at $t = 0$ so that for some time period $x_1(t) \geqslant M_1$. During this time period

$$\frac{dx_1}{dt} = -a_1 x_1 x_2 + b_1 R_1 e^{-k_1 t},$$

$$\frac{dx_2}{dt} = a_3(x_1 - M_1) - a_4 x_2.$$

Now these nonlinear equations can be approximated using the variable changes

$$x_1 = M_1 + z_1,$$
$$x_2 = 0 + z_2,$$

by the linear equations

$$\frac{dz_1}{dt} = -a_1 M_1 z_2 + b_1 R_1 e^{-k_1 t},$$

$$\frac{dz_2}{dt} = a_3 z_1 - a_4 z_2,$$

where

$$z_1(0) = 0, \quad z_2(0) = 0.$$

Taking the Laplace transform of these equations obtain

$$Z_1 = \frac{(s + a_4)b_1 R_1}{(s + k)d(s)},$$

$$Z_2 = \frac{a_3 b_1 R_1}{(s + k)d(s)},$$

where

$$d(s) = s^2 + a_4 s + a_3 a_1 M_1.$$

Since all constants are positive this is a stable characteristic polynomial. Depending on the values of the constants, the response for z_1 may or may not remain positive. Only if it remains positive is the above linearized model valid. In any event, $z_1(t) \to 0$ for some finite value of t or as $t \to \infty$. Thus $x_1(t) \to M_1$ for some finite value of t or as $t \to \infty$. Even if for some value of t, $x_1(t) < M_1$, an analysis of the model in this case, using a linearization approach, leads to the same conclusion that $x_1(t) \to M_1$ for some finite value of t or as $t \to \infty$ and that $x_2(t) \to 0$ as $t \to \infty$. Hence, the qualitative fact emerges that $x_1(t) \to M_1$ and $x_2(t) \to 0$ as $t \to \infty$. A typical response is indicated in Figure 6-5-1.

The responses indicated qualitatively above, assuming all coefficients in the model are positive, are characteristic of a normal individual. In a

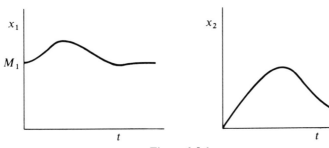

Figure 6-5-1

diabetic individual insulin is not secreted by the pancreas when there is an excess blood sugar level. Thus the coefficient a_3 in the model is very small; for explanatory purposes suppose that $a_3 = 0$. As indicated above, consider

$$u_1(t) = R_1 e^{-k_1 t}, \quad u_2(t) = 0, \quad t \geqslant 0$$

and

$$x_1(0) = M_1, \quad x_2(0) = 0.$$

Then the variational variable z_1 has the Laplace transform

$$Z_1 = \frac{(s + a_4) b_1 R_1}{(s + k_1) d(s)},$$

where $d(s) = s^2 + a_4 s$ since $a_3 = 0$ in this case. Thus the characteristic polynomial is not stable. Using the final value theorem it follows that $z_1(t) \to b_1 R_1 / k_1$ as $t \to \infty$; hence $z_1(t) > 0$ and thus $x_1(t) > M_1$ for $t \geqslant 0$ so that necessarily $x_1(t) \to M_1 + b_1 R_1 / k_1$. Note also that $x_2(t) = 0$, $t \geqslant 0$. Hence, as indicated in Figure 6-5-2, the blood sugar level for a diabetic individual is not properly regulated by the pancreas.

For a diabetic individual, where $a_3 = 0$, consider now an insulin input with

$$u_1(t) = 0, \quad u_2(t) = R_2 e^{-k_2 t}, \quad t \geqslant 0,$$

supposing an initially high blood sugar level with

$$x_1(0) > M_1, \quad x_2(0) = 0.$$

In this case, assuming that $x_1(t) \geqslant M_1$ the relevant equations are

$$\frac{dx_1}{dt} = -a_1 x_1 x_2,$$

$$\frac{dx_2}{dt} = -a_4 x_2 + b_2 R_2 e^{-k_2 t}.$$

Making the variable change

$$x_1 = M_1 + z_1,$$

$$x_2 = 0 + z_2$$

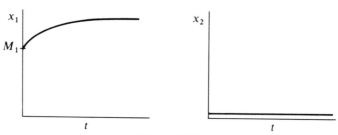

Figure 6-5-2

leads to the linearized model

$$\frac{dz_1}{dt} = -(a_1 M_1)z_2,$$

$$\frac{dz_2}{dt} = -a_4 z_2 + b_2 R_2 e^{-k_2 t}$$

with $z_1(0) = x_1(0) - M_1$, $z_2(0) = 0$. Taking Laplace transforms obtain

$$Z_1 = \frac{x_1(0) - M_1}{s} - \frac{a_1 M_1 b_2 R_2}{s(s + k_2)(s + a_4)},$$

$$Z_2 = \frac{b_2 R_2}{(s + k_2)(s + a_4)}.$$

Assuming that $k_2 \neq a_4$, then as $t \to \infty$,

$$z_1(t) \to x_1(0) - M_1 - \frac{a_1 M_1 b_2 R_2}{k_2 a_4},$$

$$z_2(t) \to 0$$

so that

$$x_1(t) \to x_1(0) - \frac{a_1 M_1 b_2 R_2}{k_2 a_4},$$

$$x_2(t) \to 0,$$

i.e., the blood sugar level tends to a constant value less than the initial blood sugar level and the insulin level tends to zero although it is not identically zero. Of course, proper insulin dosage to achieve the condition that $x_1(t) \to M_1$ as $t \to \infty$ is achieved by choosing the insulin input constants R_2 and k_2 to satisfy

$$M_1 = x_1(0) - \frac{a_1 M_1 b_2 R_2}{k_2 a_4}$$

which leads to

$$\frac{R_2}{k_2} = \frac{a_4}{a_1 b_2}\left[\frac{x_1(0)}{M_1} - 1\right].$$

With such a choice the time responses are indicated in Figure 6-5-3.

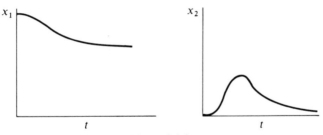

Figure 6-5-3

It should be clear that additional analysis could be carried out using the model that has been developed. In particular, the responses of a diabetic to sequential food and insulin inputs over a several day period could be analyzed; of course one objective of such an analysis might be to determine the levels of insulin input to achieve good regulation of the diabetic's blood sugar level near the equilibrium value.

The results of a computer simulation study are given in Appendix E.

Exercises

6-1. Consider the two dimensional motion of a boat as described in Case Study 6-1 with the parameter values

$$W = 2.0 \text{ miles/hr}, \quad V = 10.0 \text{ miles/hr}, \quad L = 1.0 \text{ mile}.$$

Assume that initially the position of the boat is given by

$$x_1(0) = 0, \quad x_2(0) = 0.$$

Consider a constant heading angle of $\alpha(t) = 10.0$ degrees for $t \geqslant 0$. How long does it take for the boat to cross the river? What is the path of the boat? What is the downstream location of the boat when the opposite shore is reached?

6-2. Consider the epidemic model described in Case Study 6-2 with the parameter values

$$\beta = 0.0001, \quad \gamma = 0.08.$$

Assume that there is no treatment available to stop the spread of the disease so that $u(t) = 0$ for $t \geqslant 0$. Assume that initially there are 9990 susceptible individuals and 10 infective individuals in a closed population of 10000.

What is the total number of individuals that become infected? Ultimately, what is the ratio of individuals that have become infected to the total population?

6-3. Consider the tunnel diode circuit as described in Case Study 6-3 with the parameter values

$$R = 500.0 \text{ ohm}, \ L = 0.01 \text{ henry}, \ C = 10^{-6} \text{ farad}$$

and with the diode characteristics given by the relation

$$i_d = 0.010 \, v_d \qquad \text{if} \quad v_d \leqslant 1.0,$$

$$i_d = 0.015 - 0.005 \, v_d \qquad \text{if} \quad 1.0 < v_d \leqslant 2.0,$$

$$i_d = -0.015 + 0.010 \, v_d \qquad \text{if} \quad v_d > 2.0,$$

where i_d is the diode current in amps and v_d is the diode voltage in volts.

(a) Suppose that the input voltage is $e(t) = 2.5$ volts for $t \geqslant 0$. What is the zero state response of the circuit?

(b) Suppose that the input voltage is $e(t) = 2.5$ volts for $t \geqslant 0$. What is the circuit response if initially $v(0) = 4.0$ volts and $i(0) = 0$ amps?

(c) Suppose that the input voltage is $e(t) = 5.0$ volts for $t \geqslant 0$. What is the zero state response of the circuit?

(d) Suppose that the input voltage is $e(t) = 5.0$ volts for $t \geqslant 0$. What is the circuit response if initially $v(0) = 4.0$ volts and $i(0) = 0$ amps?

(e) Suppose that the input voltage is $e(t) = 10.0$ volts for $t \geqslant 0$. What is the zero state response of the circuit?

(f) Suppose that the input voltage is $e(t) = 10.0$ volts for $t \geqslant 0$. What is the circuit response if initially $v(0) = 4.0$ volts and $i(0) = 0$ amps?

6-4. Consider the underwater launch of a rocket as described in Case Study 6-4 with the parameter values

$$W = 100.0 \text{ lb}, \quad \mu = 2.5 \text{ (lb-sec)}/\text{ft}.$$

(a) What is the maximum altitude of the rocket if there is no thrust, i.e., $T(t) = 0$ for $t \geqslant 0$ and initially

$$h(0) = -100.0 \text{ ft}, \quad \frac{dh}{dt}(0) = 5000.0 \text{ ft}/\text{sec}?$$

How long does it take for the rocket to reach its maximum altitude?

(b) What is the maximum altitude of the rocket if the thrust on the rocket is given by

$$T(t) = 5000.0 \text{ lb}, \quad 0 \leqslant t < 2.0 \text{ sec},$$
$$T(t) = 0.0 \text{ lb}, \quad t \geqslant 2.0 \text{ sec},$$

and initially

$$h(0) = -100.0 \text{ ft}, \quad \frac{dh}{dt}(0) = 0.0 \text{ ft}/\text{sec}?$$

How long does it take for the rocket to reach its maximum altitude?

6.5. A model for blood sugar and insulin levels was developed in Case Study 6-5. Suppose that the blood sugar and insulin levels are each measured in the units of milligrams per cubic centimeter. Assume the parameter values in the model are

$$a_2 = 1.0 \text{ 1}/\text{hr}, \quad b_1 = 1.0 \text{ 1}/\text{cc},$$
$$a_4 = 2.0 \text{ 1}/\text{hr}, \quad M_1 = 100.0 \text{ mg}/\text{cc}.$$

Suppose that there is no insulin input and that the sugar input, corresponding to three meals, is an indicated below.

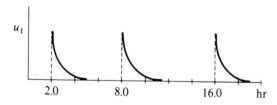

Each "spike" is exponential in form with a magnitude of 100 mg/hr and time constant of 0.5 hr. Initially the blood sugar and insulin levels are

$$x_1(0) = 100.0 \text{ mg/cc}, \quad x_2(0) = 0.0 \text{ mg/cc}.$$

(a) For a normal individual suppose that

$$a_1 = 0.05 \text{ cc/(hr-mg)}, \quad a_3 = 0.5 \text{ 1/hr}.$$

What are the blood sugar and insulin levels for $t \geqslant 0$ in such an individual?

(b) For a diabetic individual suppose that

$$a_1 = 0.05 \text{ cc/(hr-mg)}, \quad a_3 = 0.01 \text{ 1/hr}.$$

What are the blood sugar and insulin levels for $t \geqslant 0$ in such an individual?

6-6. Consider the two dimensional motion of a boat in a river as described in Case Study 6-1. The notation and assumptions of that case study are assumed except that the velocity profile of the river current is assumed to be

$$\frac{W}{2}\left[1 - \cos\frac{2\pi x_1}{L}\right] \quad \text{for} \quad 0 \leqslant x_1 \leqslant L.$$

Assume that the parameter values are

$$W = 2.0 \text{ miles/hr}, \quad V = 10.0 \text{ miles/hr}, \quad L = 1.0 \text{ mile}$$

and assume that initially

$$x_1(0) = 0.0, \quad x_2(0) = 0.0.$$

(a) Develop a state model for the motion of the boat.
(b) Suppose that the boat is moving with a constant heading angle $\alpha(t) = 10.0$ degrees for $t \geqslant 0$. How long does it take for the boat to cross the river? What is the path of the boat? What is the downstream location of the boat when the opposite shore is reached?

6-7. A tank initially contains 10.0 liters of salt solution and to it is added a salt solution containing 5.0 grams of salt per liter, pumped into the tank at the constant rate of 5.0 liters per minute. The solution in the tank is thoroughly mixed and pumped out of the tank at the constant rate of 8.0 liters per minute.

(a) Choose state variables and develop a state model.
(b) What is the concentration of salt in the tank for $t \geqslant 0$?
(c) What is the maximum concentration of salt in the tank? When does this occur?
(d) How long does it take for the tank to empty?

6-8. Consider an LRC circuit containing an electronic device; the circuit is indicated below with a resistance of 1.0 ohm, an inductance of 0.1 henry and capacitance of 10^{-6} farad.

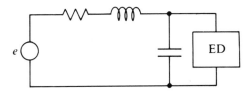

Assume the ideal voltage source e is the input and the voltage drop v_d across the electronic device is the output. Denote the current through the electronic device by i_d; then the current-voltage relation for the electronic device is

$$i_d = v_d + 1.0 \quad \text{if} \quad v_d \geqslant 0,$$
$$i_d = v_d - 1.0 \quad \text{if} \quad v_d < 0$$

with i_d measured in amps and v_d measured in volts.

(a) Choose state variables and write state equations.
(b) If the input voltage is constant 2.0 volts for $t \geqslant 0$ what are the equilibrium states?
(c) Which equilibrium states are stable?
(d) For each equilibrium state develop a linearized set of state equations assuming the state values are near the equilibrium values.

6-9. Consider the vertical ascent of a rocket from the Earth. Let m denote the constant mass of the rocket, let r denote the distance of the rocket from the center of the Earth, let T denote the thrust of the rocket; R is the radius of the Earth. Assume that the only forces on the rocket are an inverse square law gravitational force plus the thrust; the vertical motion of the rocket is described by

$$m\frac{d^2r}{dt^2} = -mg\left(\frac{R}{r}\right)^2 + T.$$

Suppose that

$$g = 32.2 \text{ ft/sec}^2, \quad R = 4000.0 \text{ miles}.$$

Consider the thrust as the input variable.

(a) Choose state variables and write state equations for the vertical motion of the rocket.

(b) Suppose there is no thrust on the rocket so that $T(t) = 0$ for $t \geq 0$. Sketch some typical trajectories in the dr/dt versus r phase plane for $r \geq R$.

(c) Suppose there is no thrust so that $T(t) = 0$ for $t \geq 0$. There is some initial velocity V at which the rocket can escape from the Earth's gravitational attraction, i.e., if $r(0) = R$ and $(dr/dt)(0) \geq V$ then $r(t) \to \infty$ as $t \to \infty$; if $r(0) = R$ and $(dr/dt)(0) < V$ the rocket ultimately returns to Earth. What is the escape velocity V?

(d) Assume that there is no thrust on the rocket and that the rocket leaves the surface of the Earth with an initial velocity which exactly equals the escape velocity. Describe the vertical motion of the rocket.

(e) Suppose that the rocket thrust exactly balances the gravitational force on the rocket, i.e., $T(t) = mg$ for $t \geq 0$. Sketch some typical trajectories in the dr/dt versus r phase plane for $r \geq R$.

(f) Suppose that $T(t) = mg$ for $t \geq 0$. Describe the vertical motion of the rocket if initially $r(0) = R$ and $(dr/dt)(0) > 0$.

Chapter VII

Higher Order Nonlinear State Models

Higher order state models are considered in this chapter. As indicated previously, it is not possible to develop completely general procedures for analyzing such models. But it is possible to indicate certain special analytical procedures that, in special cases, may be of aid in determining the response properties of a dynamic system.

Some Systems Theory

A set of nonlinear equations of the form

$$\frac{dx_1}{dt} = f_1(x_1, \ldots, x_n, u),$$

$$\vdots$$

$$\frac{dx_n}{dt} = f_n(x_1, \ldots, x_n, u),$$

$$y = g(x_1, \ldots, x_n, u)$$

defines the state model of interest in this chapter. The state $x_1, \ldots, x_n$ is an n-tuple of variables; as usual u is the input variable and y is the output variable. The model is defined by the $n + 1$ functions $f_1, \ldots, f_n$ and g.

As indicated in the previous two sections it is not usually possible to determine the general response as a function of the initial state and the input function; the general response can be determined only in a few cases, for example, if many of the equations are uncoupled, i.e., the function f_i

depends only on the variable x_i and u, for some i. General solution procedures are not pursued further here.

Of course it is possible to determine a special class of responses, namely the constant responses. Suppose that the input is constant $u(t) = \bar{u}$, $t \geqslant 0$. Then if for some values $\bar{x}_1, \ldots, \bar{x}_n$,

$$0 = f_1(\bar{x}_1, \ldots, \bar{x}_n, \bar{u}),$$

$$\vdots$$

$$0 = f_n(\bar{x}_1, \ldots, \bar{x}_n, \bar{u}),$$

it follows that if $x_1(0) = \bar{x}_1, \ldots, x_n(0) = \bar{x}_n$ then necessarily $x_1(t) = \bar{x}_1, \ldots, \bar{x}_n(t) = \bar{x}_n$ for all $t \geqslant 0$, i.e., the state $\bar{x}_1, \ldots, \bar{x}_n$ is an equilibrium state corresponding to the constant input $\bar{u}$. The corresponding output function is necessarily constant and is given by $y(t) = g(\bar{x}_1, \ldots, \bar{x}_n, \bar{u})$ for all $t \geqslant 0$. Thus the equilibrium states can be determined by solving a set of nonlinear algebraic equations.

Again, a procedure for approximating the nonlinear state model by a linearized state model can be presented. Assume, as above, that $\bar{x}_1, \ldots, \bar{x}_n$ is an equilibrium state corresponding to the constant input $\bar{u}$. Define new variables $z_1, \ldots, z_n$ and v and w from

$$u = \bar{u} + v,$$

$$y = g(\bar{x}_1, \ldots, \bar{x}_n, \bar{u}) + w,$$

$$x_1 = \bar{x}_1 + z_1,$$

$$\vdots$$

$$x_n = \bar{x}_n + z_n.$$

Then the nonlinear state model is equivalent to

$$\frac{dz_1}{dt} = f_1(\bar{x}_1 + z_1, \ldots, \bar{x}_n + z_n, \bar{u} + v),$$

$$\vdots$$

$$\frac{dz_n}{dt} = f_n(\bar{x}_1 + z_1, \ldots, \bar{x}_n + z_n, \bar{u} + v),$$

$$w = g(\bar{x}_1 + z_1, \ldots, \bar{x}_n + z_n, \bar{u} + v) - g(\bar{x}_1, \ldots, \bar{x}_n, \bar{u}),$$

where v and w are new input and output variables and $z_1, \ldots, z_n$ are new state variables. Each of the above nonlinear functions can be approximated by a linear function in the variables $z_1, \ldots, z_n, v$ near $z_1 = 0$, $\ldots, z_n = 0$, $v = 0$ using the approach discussed in Appendix D. The result is that

$$\frac{dz_1}{dt} = \sum_{j=1}^{n} \left(\frac{\partial f_1}{\partial x_j} \right) z_j + \left(\frac{\partial f_1}{\partial u} \right) v,$$

$$\vdots$$

$$\frac{dz_n}{dt} = \sum_{j=1}^{n} \left(\frac{\partial f_n}{\partial x_j} \right) z_j + \left(\frac{\partial f_n}{\partial u} \right) v,$$

$$w = \sum_{j=1}^{n} \left(\frac{\partial g}{\partial x_j} \right) z_j + \left(\frac{\partial g}{\partial u} \right) v.$$

All of the above partial derivatives are evaluated at $\bar{x}_1, \ldots, \bar{x}_n$ and $\bar{u}$; note that the constant terms satisfy $f_i(\bar{x}_1, \ldots, \bar{x}_n, \bar{u}) = 0$ for all $i = 1, \ldots, n$ since $\bar{x}_1, \ldots, \bar{x}_n$ is an equilibrium state with constant input $\bar{u}$. The linearized model above can thus be considered to be an approximation to the nonlinear state model under the assumption that each x_i is sufficiently close to $\bar{x}_i$ and u is sufficiently close to $\bar{u}$, i.e., $z_1, \ldots, z_n$ and v are all sufficiently small. Of course, the obvious advantage of the linearized model is that it is relatively easy to determine the responses on the basis of the linearized model, but it is important to keep in mind that the model is valid and leads to correct results only if the stated assumptions are satisfied.

Of course there are numerous other special analytical techniques which may be of use in the analysis of special state models. A few of these special techniques are illustrated in the case study examples in the following sections.

Case Study 7-1: Continuous Flow Stirred Tank Chemical Reactor

One of the most common devices in the chemical industry is a continuous flow stirred tank reactor; this is a device which allows certain chemical reactions to proceed under controlled circumstances. In this example a rather simple version of a reactor is considered but the important features are considered. As the name implies a continuous flow stirred tank reactor allows continuous inflow of certain reactants into a tank; as the reactants enter the tank certain chemical reactions occur; the mixture in the tank is kept completely mixed and appropriate amount of the mixture is drained off. In this particular example, it is assumed that H_2O and SO_3 are added to the tank and and that they undergo an irreversible chemical reaction to form sulfuric acid H_2SO_4 according to the chemical equation

$$H_2O + SO_3 \rightarrow H_2SO_4,$$

where one mole of H_2O and one mole of SO_3 react to form one mole of H_2SO_4. Assume that the tank contains a fixed volume V and that the temperature inside the tank is a constant. Let Q denote the constant flow rate into the tank and out of the tank. The inflow is assumed to contain a molar concentration C_A of H_2O and a molar concentration C_B of SO_3; the inflow is assumed not to contain any H_2SO_4. The outflow is assumed to consist of the mixture within the tank, assumed to be uniformly and completely mixed. For purposes of the present example there are two input variables, namely the two concentrations C_A and C_B in the inflow; the concentrations of the outflow are considered to be the output.

Let c_A, c_B, c_C denote the molar concentrations of H_2O, SO_3, H_2SO_4 within the tank. The rate balance equations are

$$V\frac{dc_A}{dt} = QC_A - Qc_A - Vr,$$

$$V\frac{dc_B}{dt} = QC_B - Qc_B - Vr,$$

$$V\frac{dc_C}{dt} = -Qc_C + Vr,$$

where r denotes the intrinsic rate of the chemical reaction. It is usual to assume that the rate of reaction is proportional to the product of the molar concentrations of the reactants, that is,

$$r = kc_A c_B$$

for a constant k.

It is clear that the concentrations c_A, c_B, c_C can serve as the state variables, so that $x_1 = c_A$, $x_2 = c_B$, $x_3 = c_C$. The two concentrations of the inflow are the input variables so that $u_1 = C_A$, $u_2 = C_B$. Consequently the state equations are the three first order nonlinear equations

$$\frac{dx_1}{dt} = -\left(\frac{Q}{V}\right)x_1 - kx_1 x_2 + \left(\frac{Q}{V}\right)u_1,$$

$$\frac{dx_2}{dt} = -\left(\frac{Q}{V}\right)x_2 - kx_1 x_2 + \left(\frac{Q}{V}\right)u_2,$$

$$\frac{dx_3}{dt} = -\left(\frac{Q}{V}\right)x_3 + kx_1 x_2.$$

An important feature of this nonlinear state model is that the first two state equations are uncoupled from the third equation, i.e., the expressions for dx_1/dt and dx_2/dt do not depend on x_3. This observation is important since the first two state equations could be analyzed as second order equations; the last state equations could subsequently be analyzed as a first order equation. In fact if $x_1(t), x_2(t) \geqslant 0$, for $t \geqslant 0$, are the concentrations of H_2O and SO_3 then it is clear that the output concentration of H_2SO_4 for $t \geqslant 0$ is

$$x_3(t) = \exp\left(\frac{-Qt}{V}\right)x_3(0) + \int_0^t \exp\left(\frac{-Q}{V}(t-\tau)\right)kx_1(\tau)x_2(\tau)\,d\tau.$$

One case of interest is where the concentrations of H_2O and SO_3 added to the tank are constant; thus the inputs are constants $u_1(t) = \bar{C}_A$ and $u_2(t) = \bar{C}_B$ for $t \geqslant 0$. Thus if there is an equilibrium state the three algebraic equations

$$0 = -\left(\frac{Q}{V}\right)x_1 - kx_1x_2 + \left(\frac{Q}{V}\right)\bar{C}_A,$$

$$0 = -\left(\frac{Q}{V}\right)x_2 - kx_1x_2 + \left(\frac{Q}{V}\right)\bar{C}_B,$$

$$0 = -\left(\frac{Q}{V}\right)x_3 + kx_1x_2$$

must be satisfied. These algebraic equations can be solved in the following manner. Using the first equation it is possible to solve for x_1 in terms of x_2 and $\bar{C}_A$; this can then be substituted into the second equation to obtain a quadratic expression in x_2, namely

$$kx_2^2 + \left[\frac{Q}{V} + k\bar{C}_A - k\bar{C}_B\right]x_2 - \left(\frac{Q}{V}\right)\bar{C}_B = 0.$$

Without solving this quadratic equation explicitly it can be recognized that there is necessarily a positive solution and a negative solution. Only the positive solution is of physical interest; suppose it is denoted by $\bar{C}_2$. Thus the corresponding value of x_1 can be determined from the first equation and the corresponding value of x_3 can then be determined from the third equation; suppose these values are given by $\bar{C}_1$ and $\bar{C}_3$. Thus $\bar{C}_1, \bar{C}_2, \bar{C}_3$ are the only positive solutions of the above algebraic equations and hence $\bar{C}_1, \bar{C}_2, \bar{C}_3$ is the only equilibrium state of physical interest.

It is now possible to obtain a linear model by assuming that the state is sufficiently close to the equilibrium state. Thus if the change of variables

$$x_1 = \bar{C}_1 + z_1,$$

$$x_2 = \bar{C}_2 + z_2,$$

$$x_3 = \bar{C}_3 + z_3,$$

$$u_1 = \bar{C}_A + v_1,$$

$$u_2 = \bar{C}_B + v_2$$

is made then the nonlinear state equations can be written as

$$\frac{dz_1}{dt} = -\left(\frac{Q}{V}\right)(\overline{C}_1 + z_1) - k(\overline{C}_1 + z_1)(\overline{C}_2 + z_2) + \left(\frac{Q}{V}\right)(\overline{C}_A + v_1),$$

$$\frac{dz_2}{dt} = -\left(\frac{Q}{V}\right)(\overline{C}_2 + z_2) - k(\overline{C}_1 + z_1)(\overline{C}_2 + z_2) + \left(\frac{Q}{V}\right)(\overline{C}_B + v_2),$$

$$\frac{dz_3}{dt} = -\left(\frac{Q}{V}\right)(\overline{C}_3 + z_3) + k(\overline{C}_1 + z_1)(\overline{C}_2 + z_2).$$

Now using the fact that $\overline{C}_1, \overline{C}_2, \overline{C}_3$ is an equilibrium state and ignoring the product terms $z_1 z_2$ in comparison with the linear terms there results the linearized state model

$$\frac{dz_1}{dt} = -\left[\frac{Q}{V} + k\overline{C}_2\right]z_1 - (k\overline{C}_1)z_2 + \left(\frac{Q}{V}\right)v_1,$$

$$\frac{dz_2}{dt} = -(k\overline{C}_2)z_1 - \left[\frac{Q}{V} + k\overline{C}_1\right]z_2 + \left(\frac{Q}{V}\right)v_2,$$

$$\frac{dz_3}{dt} = (k\overline{C}_2)z_1 + (k\overline{C}_1)z_2 - \left(\frac{Q}{V}\right)z_3,$$

which is valid so long as z_1 and z_2 are sufficiently small; note that it is not required that x_3 is near $\overline{C}_3$. After some calculations it can be shown that for this linearized model the characteristic polynomial is

$$d(s) = \left[s + \frac{Q}{V}\right]\left[s^2 + \left(\frac{2Q}{V} + k\overline{C}_2 + k\overline{C}_1\right)s\right.$$

$$\left. + \left(\frac{Q}{V}\right)^2 + \left(\frac{Q}{V}\right)(k\overline{C}_2 + k\overline{C}_1) + k^2\overline{C}_1\overline{C}_2\right].$$

Thus the characteristic zeros can easily be determined for given parameter values; clearly the characteristic zeros all have negative real part so that the linearized model is stable. Thus if $u_1(t) = \overline{C}_A$ and $u_2(t) = \overline{C}_B$ for $t \geqslant 0$ and if $x_1(0)$ is near $\overline{C}_1$ and $x_2(0)$ is near $\overline{C}_2$ and for arbitrary $x_3(0)$ it follows that $x_1(t) \to \overline{C}_1$, $x_2(t) \to \overline{C}_2$ and $x_3(t) \to \overline{C}_3$ as $t \to \infty$. The equilibrium state $\overline{C}_1, \overline{C}_2, \overline{C}_3$ is said to be stable. Thus if the chemical reactor is in steady operation with constant concentrations of H_2O and SO_3 in the input flows then the reactor should produce H_2SO_4 so that its concentration in the output flow is asymptotically constant. Further, the reactor is not sensitive to disturbances; such effects will eventually be attenuated.

Our analysis has been with regard to the case where the input concentrations are constants; if these input concentrations were specific time functions, say when the reactor is shut down, an analysis might be based on the above linearized model to determine the corresponding responses.

Case Study 7-2: A Hot Air Balloon

In this example the vertical motion of a flexible spherical balloon filled with hot air is considered. The vertical motion is controlled by a heater element which supplies heat to the air inside the balloon causing the air inside the ballon to become less dense so that the balloon necessarily tends upward. It is assumed that the air inside the balloon behaves as an ideal gas so that the density of air inside the balloon is inversely proportional to the absolute temperature of the air in the balloon, assuming a constant balloon volume. Heat is added to the balloon by the heater element but heat is lost to the surrounding air at a rate proportional to the difference between the temperature of the air in the balloon and the temperature of the ambient air. If the temperature of the air in the balloon exceeds the ambient air temperature then the density of air in the balloon is smaller than the ambient air density. Consequently, there is an upward buoyant force on the balloon. If this buoyant force exceeds the weight of the balloon then the balloon will tend to move upward. As the balloon moves vertically there is an aerodynamic drag force which is assumed to be proportional to the velocity of the balloon. Pressure and density variations of the ambient air as a function of altitude are ignored. Only the vertical motion of the balloon is considered; horizontal motion is completely ignored in the analysis. The input is the rate at which heat is added to the air in the balloon by the heater element; the output is the altitude of the balloon. The basic equations which characterize the vertical motion of the balloon are now derived.

Let T denote the absolute temperature of the air inside the balloon and let w denote the weight density of the air inside the balloon. If T_a and w_a are the temperature and weight density of ambient air surrounding the balloon, both assumed constant, then the thermodynamic relation

$$wT = w_a T_a$$

follows from the ideal gas law; the volume V of air inside the balloon is assumed constant. If q denotes the heat transfer rate to the air in the balloon from the heater element then, based on the stated assumptions,

$$\frac{dT}{dt} = -\frac{1}{\tau}(T - T_a) + Kq,$$

where τ is the thermal time constant of the balloon and K depends on the thermal capacity of the balloon.

Let h denote the altitude of the balloon above sea level. The vertical motion of the balloon is described by the equation

$$\frac{W}{g}\frac{d^2h}{dt^2} = V(w_a - w) - W - \mu\frac{dh}{dt},$$

where W is the weight of the balloon, air in the balloon and the payload, g is the acceleration of gravity and μ is a drag coefficient. Thus the above expression gives the mass of the balloon multiplied by its vertical acceleration in terms of the forces on the balloon. The first term on the right hand side of the equation is the buoyant force on the balloon, the second term is the weight of the balloon and the third term is the aerodynamic drag force on the balloon. After appropriate simplification and expressing the density w in terms of the temperature T of the air in the balloon obtain

$$\frac{d^2h}{dt^2} = g\left[\frac{W_a}{W}\left(1 - \frac{T_a}{T}\right) - 1 - \frac{\mu}{W}\frac{dh}{dt}\right],$$

where the new constants $W_a = w_a V$ can be thought of as the maximum possible buoyant force on the balloon, corresponding to a vacuum inside the balloon. Now by choosing as state variables the temperature of air in the balloon and the altitude and vertical velocity of the balloon let $x_1 = T$, $x_2 = h$, $x_3 = dh/dt$. The input variable u is the heat transfer rate g supplied to the balloon by the heater. Thus the third order nonlinear state model is obtained:

$$\frac{dx_1}{dt} = -\frac{1}{\tau}(x_1 - T_a) + Ku,$$

$$\frac{dx_2}{dt} = x_3,$$

$$\frac{dx_3}{dt} = g\left[\frac{W_a}{W}\left(1 - \frac{T_a}{x_1}\right) - 1 - \frac{\mu}{W}x_3\right].$$

It is possible to solve these equations sequentially given a specified input function and initial state. The equation for x_1 is uncoupled and can be solved by itself as a first order linear equation. Given x_1 as a time function the last two equations in x_2 and x_3 are linear and can be solved. This approach to obtaining the exact response functions leads to difficult integrations; the qualitative features of the responses are somewhat obscured so the approach in not pursued.

First consider if there are equilibrium states, that is if there are any constant responses of the system. If there are equilibrium states then necessarily

$$0 = -\frac{1}{\tau}(x_1 - T_a) + Ku,$$

$$0 = x_3,$$

$$0 = \frac{W_a}{W}\left(1 - \frac{T_a}{x_1}\right) - 1 - \frac{\mu}{W}x_3.$$

These nonlinear algebraic equations have no meaningful solution if

$W_a < W$. In this case the buoyant force can never be sufficient to maintain the balloon in equilibrium. Hence it is assumed in what follows that $W_a > W$. Then the above algebraic equations have the solution

$$x_1 = \overline{T} = \left(\frac{W_a}{W_a - W} \right) T_a,$$

$$x_2 = \overline{H},$$

$$x_3 = 0,$$

if

$$u = \overline{Q} = \frac{T_a}{K\tau} \left(\frac{W}{W_a - W} \right),$$

where $\overline{T}, \overline{Q}$ are constants defined above and $\overline{H}$ is an arbitrary constant. Thus there is an equilibrium state only if the input is given by $u(t) = \overline{Q}$ for $t \geq 0$, in which case the equilibrium state is $\overline{T}, \overline{H}, 0$, where $\overline{T}$ is given above and $\overline{H}$ is completely arbitrary. Thus the balloon can be in equilibrium only if the heater output is given by $\overline{Q}$ above; equilibrium can occur at any altitude with the corresponding temperature $\overline{T}$ given above and zero vertical velocity. Note that $\overline{Q} > 0$ and $\overline{T} > T_a$.

If the balloon is not too far from its equilibrium state and the input is not too different from the value $\overline{Q}$ then it is possible to approximate the above nonlinear model by a linearized state model. Thus define the variations z_1, z_2, z_3 and v by

$$x_1 = \overline{T} + z_1,$$

$$x_2 = \overline{H} + z_2,$$

$$x_3 = 0 + z_3,$$

$$u = \overline{Q} + v,$$

where $\overline{T}$ and $\overline{Q}$ are given above and $\overline{H}$ is an arbitrary but fixed value. Then in terms of these variation variables

$$\frac{dz_1}{dt} = -\frac{1}{\tau} z_1 + Kv,$$

$$\frac{dz_2}{dt} = z_3,$$

$$\frac{dz_3}{dt} = g \left[\frac{W_a}{W} \left(1 - \frac{T_a}{\overline{T} + z_1} \right) - 1 - \frac{\mu}{W} z_3 \right].$$

Now, assuming that z_1 is sufficiently small the nonlinear term $(\overline{T} + z_1)^{-1}$ can be approximated by the linear terms in z_1, namely $(\overline{T})^{-1} - (\overline{T})^{-2} z_1$.

Using this approximation and substituting the expression for $\bar{T}$ obtain

$$\frac{dz_1}{dt} = -\frac{1}{\tau}z_1 + Kv,$$

$$\frac{dz_2}{dt} = z_3,$$

$$\frac{dz_3}{dt} = g\left[\frac{(W_a - W)^2}{W_a W}\frac{z_1}{T_a} - \frac{\mu}{W}z_3\right].$$

Note that this linearized model is such that the first state equation is uncoupled from the last two state equations. It is an easy matter to determine the transfer function relations

$$\frac{Z_1}{V} = \frac{K\tau}{1 + \tau s},$$

$$\frac{Z_2}{V} = \frac{g(W_a - W)^2(K\tau/T_a)}{W_a Ws(1 + \tau s)(s + \mu g/W)},$$

$$\frac{Z_3}{V} = \frac{g(W_a - W)^2(K\tau/T_a)}{W_a W(1 + \tau s)(s + \mu g/W)}.$$

Actually these expressions are sufficiently simple that the general response, based on the linear state model, can be analytically determined. But rather than obtain these general responses the qualitative responses in some special cases are determined.

Suppose that $x_1(0) = (W_a/(W_a - W))T_a$, $x_2(0) = \bar{H}$, $x_3(0) = 0$ where $\bar{H}$ is arbitrary; determine the response for a constant heater input $u(t) = T_a(W/(W_a - W))/K\tau + q_0$, $t \geq 0$, where the constant q_0 is small compared to $\bar{Q}$. In terms of the variations, clearly $z_1(0) = 0$, $z_2(0) = 0$, $z_3(0) = 0$ and $v(t) = q_0$, $t \geq 0$. Thus the linearized model can be used to determine the resulting response. Without going into the details it is easily determined, using the transfer functions previously derived, that in this case as $t \to \infty$

$$z_1(t) \to K\tau q_0,$$

$$z_2(t) \to \frac{(W_a - W)^2 K\tau q_0}{W_a \mu T_a}\left[t - \left(\frac{W}{\mu g} + \tau\right)\right],$$

$$z_3(t) \to \frac{(W_a - W)^2 K\tau q_0}{W_a \mu T_a}.$$

Consequently, in terms of the earlier variables, as $t \to \infty$ it follows that

$$x_1(t) \to \left(\frac{W_a}{W_a - W} \right) T_a + K\tau q_0,$$

$$x_2(t) \to H + \frac{(W_a - W)^2 K\tau q_0}{W_a \mu T_a} \left[t - \left(\frac{W}{\mu g} + \tau \right) \right],$$

$$x_3(t) \to \frac{(W_a - W)^2 K\tau q_0}{W_a \mu T_a}.$$

One clear observation is that if $q_0 > 0$ then the altitude of the balloon increases and the temperature of air in the balloon increases; if $q_0 < 0$ then the altitude and temperature decrease. Thus the altitude of the balloon can be controlled by suitable choice of the heat added to the balloon from the heater element. The details of the transient response are determined by the poles of the transfer functions given above. Qualitative time responses in the case $q_0 > 0$ are indicated in Figure 7-2-1. If $q_0 < 0$ then the time responses are as indicated in Figure 7-2-2.

The model that has been developed has been used only for the analysis of some very simple manuevers; the model could be used to analyze numerous other situations as well, e.g., achieving a soft landing, climbing to a certain altitude to reach equilibrium, etc. The basic nonlinear model characterizes the resulting motions in each case; in most cases the linearized model can be used as well.

A simple computer simulation for this case study example is given in Appendix E.

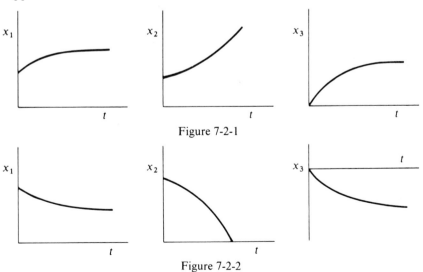

Figure 7-2-1

Figure 7-2-2

Case Study 7-3: Motion of a Rocket Near Earth

In this case study the motion of a rocket in the Earth's gravitational field is studied; this is a special case of the "two-body" problem. The Earth is assumed to be fixed in space and the motion of the rocket, considered a point mass, is examined. The rocket is assumed to be outside of the Earth's atmosphere so that the only forces acting on the rocket are the gravitational force of the Earth and the propulsive force of the rocket. The gravitational force F is given by the usual inverse square relation

$$F = -mg \frac{R^2}{r^2},$$

where r is the distance of the rocket from the center of the Earth, R is the radius of the Earth and g is the acceleration of gravity at the surface of the Earth. The propulsive force on the rocket is given by the thrust T; the thrust T and the mass m of the rocket are assumed to be constant. As indicated in Figure 7-3-1 the thrust angle α is the angle between the thrust vector and the local vertical.

Considering only motion of the rocket within a fixed plane the position of the rocket is determined, with respect to a coordinate system fixed in space, by the polar coordinates r and θ, or equivalently by the rectangular coordinates x and y.

The equations of motion of the rocket are now derived; note that as the rocket moves the coordinates of the rocket r and θ, or equivalently x and y, change with time. The acceleration of the rocket is first determined. Clearly

$$x = r\cos\theta, \quad y = r\sin\theta$$

so that

$$\dot{x} = \dot{r}\cos\theta - r\dot{\theta}\sin\theta,$$
$$\dot{y} = \dot{r}\sin\theta + r\dot{\theta}\cos\theta,$$

where the dot notation indicates differentiation with respect to time. Differentiating again,

$$\ddot{x} = (\ddot{r} - r\dot{\theta}^2)\cos\theta - (r\ddot{\theta} + 2\dot{r}\dot{\theta})\sin\theta,$$
$$\ddot{y} = (\ddot{r} - r\dot{\theta}^2)\sin\theta + (r\ddot{\theta} + 2\dot{r}\dot{\theta})\cos\theta.$$

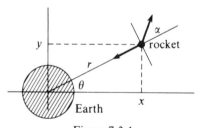

Figure 7-3-1

If a_r denotes the acceleration of the rocket in the direction of the local vertical then

$$a_r = \ddot{x}\cos\theta + \ddot{y}\sin\theta = \ddot{r} - r\dot{\theta}^2;$$

if a_T denotes the acceleration of the rocket in the local horizontal direction then

$$a_T = \ddot{y}\cos\theta - \ddot{x}\sin\theta = r\ddot{\theta} + 2\dot{r}\dot{\theta}.$$

The equations of motion of the rocket can now be obtained using Newton's laws. In the direction of the local vertical at the rocket

$$ma_r = -\frac{mgR^2}{r^2} + T\cos\alpha$$

and in the direction of the local horizontal

$$ma_T = T\sin\alpha.$$

Substituting for a_r and a_T obtain the equations of motion

$$m\left[\frac{d^2r}{dt^2} - r\left(\frac{d\theta}{dt}\right)^2\right] = -\frac{mgR^2}{r^2} + T\cos\alpha,$$

$$m\left[r\frac{d^2\theta}{dt^2} + 2\frac{dr}{dt}\frac{d\theta}{dt}\right] = T\sin\alpha.$$

Of course these equations could be expressed in terms of the rectangular coordinates x and y of the rocket, but the above equations in polar form are a bit simpler.

There are a number of possible choices for the state variables, but it should be clear that four state variables are required. The simple choice of state variables given by $x_1 = r$, $x_2 = dr/dt$, $x_3 = \theta$, $x_4 = d\theta/dt$ is made here; the input variable u is the angle α. Thus the fourth order nonlinear state model is obtained:

$$\frac{dx_1}{dt} = x_2,$$

$$\frac{dx_2}{dt} = x_1 x_4^2 - \frac{gR^2}{x_1^2} + \frac{T}{m}\cos u,$$

$$\frac{dx_3}{dt} = x_4,$$

$$\frac{dx_4}{dt} = -\frac{2x_2 x_4}{x_1} + \frac{T}{m}\sin u.$$

Clearly the model is valid only for $x_1 \geqslant R$.

This state model could be used to analyze a number of different types of rocket motion. Here this state model is used to study orbiting motions of the rocket as it moves around the Earth.

First suppose the thrust to mass ratio $T/m = 0$; there are a number of different orbits which are possible, depending on the initial state. For example, there are parabolic, elliptic and hyperbolic orbits. In this analysis the simplest case of a circular orbit is considered. If $T/m = 0$, as is assumed, and if the initial state is $x_1(0) = \rho$, $x_2(0) = 0$, $x_3(0) = 0$, $x_4(0) = \sqrt{gR^2/\rho^3}$, where $\rho > R$ is an arbitrary constant, then it is easily seen that the resulting response for $t \geqslant 0$ is

$$x_1(t) = \rho,$$

$$x_2(t) = 0,$$

$$x_3(t) = \sqrt{\frac{gR^2}{\rho^3}}\; t,$$

$$x_4(t) = \sqrt{\frac{gR^2}{\rho^3}}\;.$$

This is indeed a response since it is easily verifed that the state equations are satisfied. This particular response corresponds to a circular orbit since the rocket moves around the Earth a constant distance from the Earth at a constant angular velocity. Note that a circular orbit is possible at any altitude but the initial angular velocity must be precisely specified so that

$$\rho\left(\frac{d\theta}{dt}(0)\right)^2 = \frac{gR^2}{\rho^2},$$

i.e., the so-called centrifugal force just balances the gravitational force. The period of the motion, i.e., the time for the rocket to make one complete revolution of the Earth, is easily determined to be

$$2\pi\sqrt{\frac{\rho^3}{gR^2}}\;.$$

If the rocket is just outside the Earth's atmosphere, say $\rho = 4000$ miles then the required angular velocity for a circular orbit is 0.72 rad/min and the period is about 87 min.

The conditions for a circular orbit are particularly important; now consider motion corresponding to a near-circular orbit; it is also assumed that the thrust to mass ratio T/m is small, but not necessarily zero. In the succeeding discussion motion near a particular circular orbit for a fixed value of $\rho > R$ is considered. The approach is to obtain a linear model which characterizes the motion of a rocket under the stated conditions. To this end first make the change of state variables

$$x_1 = \rho + z_1,$$

$$x_2 = 0 + z_2,$$

$$x_3 = \omega t + z_3,$$

$$x_4 = \omega + z_4,$$

where for simplicity define $\omega^2 = gR^2/\rho^3$. State equations expressed in terms of the variables z_1, z_2, z_3, z_4 are easily obtained:

$$\frac{dz_1}{dt} = z_2,$$

$$\frac{dz_2}{dt} = (\rho + z_1)(\omega + z_4)^2 - \frac{gR^2}{(\rho + z_1)^2} + \frac{T}{m}\cos u,$$

$$\frac{dz_3}{dt} = z_4,$$

$$\frac{dz_4}{dt} = \frac{-2z_2}{(\rho + z_1)}(\omega + z_4) + \frac{T}{m(\rho + z_1)}\sin u.$$

Now assuming that the thrust to mass ratio is small and the motion of the rocket is near a circular orbit the above nonlinear functions in the variables $T/m, z_1, z_2, z_3, z_4$ can be linearized about the values $T/m = 0$, $z_1 = 0$, $z_2 = 0$, $z_3 = 0$, $z_4 = 0$ to obtain the linearized equations which follow. Note that since the thrust to mass ratio is assumed to be small it is included in the variables of interest in the linearization scheme.

$$\frac{dz_1}{dt} = z_2,$$

$$\frac{dz_2}{dt} = 3\omega^2 z_1 + 2\rho\omega z_4 + \frac{T}{m}\cos u,$$

$$\frac{dz_3}{dt} = z_4,$$

$$\frac{dz_4}{dt} = -\frac{2\omega}{\rho}z_2 + \frac{T}{m\rho}\sin u.$$

This fourth order linear state model is then an approximation to the original nonlinear state model and it should be valid so long as the motion of the rocket is near the specified circular orbit. Note, however, that the variable z_3 need not be small since no terms were ignored which included this variable.

This linearized model is now used as a basis for determining certain kinds of rocket motion. In particular consider the motion of a rocket if it is initially in a circular orbit, but there is a constant thrust T at an angle $u(t)$ from the local vertical. Thus, it is desired to determine the response corresponding to

$$x_1(0) = \rho, \quad x_2(0) = 0, \quad x_3(0) = 0, \quad x_4(0) = \omega$$

for a fixed $\rho > R$, with $u(t)$, $t \geqslant 0$ specified. Then in terms of the variation variables

$$z_1(0) = 0, \quad z_2(0) = 0, \quad z_3(0) = 0, \quad z_4(0) = 0$$

so that interest is in the zero state response of the above linear model. For convenience define the functions

$$f_1(t) = \cos u(t), \quad f_2(t) = \sin u(t), \quad t \geq 0.$$

Then, taking the Laplace transforms of the linear equations results in

$$sZ_1 = Z_2,$$

$$sZ_2 = 3\omega^2 Z_1 + \frac{T}{m} F_1,$$

$$sZ_3 = Z_4,$$

$$sZ_4 = -\frac{2\omega}{\rho} Z_2 + \frac{T}{m\rho} F_2,$$

where F_1 and F_2 are the transforms of f_1 and f_2, respectively. These equations can be solved for the transforms Z_1 and Z_2 to obtain

$$Z_1 = \frac{T}{m} \left[\frac{F_1}{s^2 + \omega^2} + \frac{2\omega F_2}{s(s^2 + \omega^2)} \right],$$

$$Z_3 = \frac{T}{m\rho} \left[\frac{-2\omega F_1}{s(s^2 + \omega^2)} + \frac{(s^2 - 3\omega) F_2}{s^2(s^2 + \omega^2)} \right].$$

First, consider the case where the thrust is always in the direction of the local vertical so that $u(t) = 0$, $t \geq 0$. Thus $f_1(t) = 1$ and $f_2(t) = 0$, $t \geq 0$. Consequently,

$$Z_1 = \frac{T}{m} \left[\frac{1}{s(s^2 + \omega^2)} \right],$$

$$Z_4 = \frac{T}{m\rho} \left[\frac{-2\omega}{s^2(s^2 + \omega^2)} \right].$$

Taking the inverse transforms obtain

$$z_1(t) = \frac{T}{m\omega^2} (1 - \cos \omega t),$$

$$z_2(t) = \frac{2T}{mg} \left[\sin \omega t - \omega t \right],$$

so that, in terms of the original state variables,

$$x_1(t) = \rho \left[1 + \frac{T}{m\omega^2} (1 - \cos \omega t) \right],$$

$$x_3(t) = \omega \left(1 - \frac{2T}{mg} \right) t + \frac{2T}{mg} \sin \omega t.$$

Thus, the motion of the rocket is an aperiodic orbit about the Earth as shown in Figure 7-3-2. The effect of a vertical thrust is to reduce the average angular velocity below that required for circular orbit; the rocket always remains within a distance $2T/m\omega^2$ of its original circular orbit. Clearly, such vertical thrust policy does not succeed in continually increasing the distance of the rocket from the Earth.

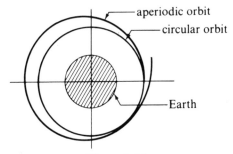

Figure 7-3-2

Next, consider the case where the thrust is always in the direction of the local horizontal so that $u(t) = \pi/2$, $t \geqslant 0$. Thus $f_1(t) = 0$ and $f_2(t) = 1$, $t \geqslant 0$. Consequently, the transforms Z_1 and Z_3 are given by

$$Z_1 = \frac{T}{m}\left[\frac{2\omega}{s^2(s^2 + \omega^2)}\right],$$

$$Z_3 = \frac{T}{m\rho}\left[\frac{(s^2 - 3\omega^2)}{s^3(s^2 + \omega^2)}\right].$$

Taking the inverse transforms obtain

$$z_1(t) = \frac{2T}{m\omega}\left(t - \frac{1}{\omega}\sin\omega t\right),$$

$$z_3(t) = \frac{T}{m\rho}\left[-\frac{3}{2}t^2 + \frac{4}{\omega^2}(1 - \cos\omega t)\right]$$

so that, in terms of the original state variables,

$$x_1(t) = \rho\left[1 + \frac{2T}{m\omega}\left(t - \frac{1}{\omega}\sin\omega t\right)\right],$$

$$x_3(t) = \omega t + \frac{T}{m\rho}\left[-\frac{3}{2}t^2 + \frac{4}{\omega^2}(1 - \cos\omega t)\right].$$

Thus as t increases the distance of the rocket from the Earth increases, so that the rocket does not remain in an orbit about the Earth. Note also that the average angular velocity is reduced below that required for circular orbit. The path of the rocket in this case is shown in Figure 7-3-3.

The linearized model has been used to examine the effects of a vertical and a horizontal thrusting policy; the resulting paths of the rocket are quite different in these two cases. The linearized equations could also be used to determine the motion for other thrust policies and for various initial states, so long as the motion is near the circular orbit so that the linearized model is a reasonable approximation to the nonlinear model.

Only a brief introduction into this problem area has been given; such analysis has been important in making progress in the exploration of space. In fact, the success of much of the space program has depended on the availability and use of good state models.

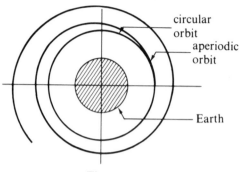

Figure 7-3-3

Case Study 7-4: Electrostatic Microphone

An important transducer which converts acoustic energy into electrical energy is an electrostatic or capacitor microphone which has a schematic diagram as shown in Figure 7-4-1. The construction of the microphone is extremely simple; a movable metal diaphragm and a fixed rigid perforated backplate are connected to an insulated frame as shown. The movable diaphragm is deflected by the pressure of a sound wave. The diaphragm and backplate form a capacitor which is biased with a constant charge. The deflection of the diaphragm changes the capacitance which in turn changes the voltage across the capacitor. As the diaphragm deflects, air is forced from the space adjacent to the diaphragm through the perforations in the backplate into the other air space. The perforations resist air flow through them, providing damping of the diaphragm motion. The air inside the air space, due to its compressibility, also creates an elastic restoring force on the diaphragm. Our interest is in determining the response characteristics of the microphone for typical input pressure waveforms; the most relevant input functions are those which are periodic or sinusoidal.

Two important variables are the charge Q on the capacitor and the displacement h of the diaphragm measured from the backplate. The equation for the electrical circuit is

$$R \frac{dQ}{dt} + \frac{Q}{C} - E = 0,$$

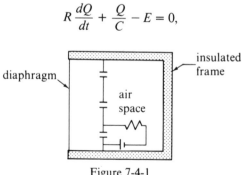

Figure 7-4-1

where R is the resistance in the circuit, E is the bias voltage in the circuit and C is the capacitance of the diaphragm and backplate. The parameters R and E are assumed constants; the capacitance C depends on the displacement of the diaphragm. The sound pressure is assumed to be uniform over the whole diaphragm exerting a total force F on the diaphragm; the diaphragm is assumed to move as a rigid body. The equation describing the motion of the diaphragm is described by

$$m\frac{d^2h}{dt^2} = -\mu\frac{dh}{dt} - k(h-L) - F_c + F,$$

where m is the mass of the diaphragm plus the mass of air accelerated by the diaphragm, μ is the damping constant and k is the elastic constant of the microphone. The parameter L denotes the position of the diaphragm when there is no force on the diaphragm. The electrostatic capacitance force on the diaphragm is denoted by F_c. It remains to describe the capacitance C and the electrostatic force F_c on the diaphragm. Our assumption is that the capacitance is inversely proportional to the spacing between the diaphragm and backplate:

$$C = \frac{B}{h}.$$

It can be shown from energy considerations that the corresponding electrostatic force on the diaphragm is given by

$$F_c = \frac{Q^2}{2B}.$$

The constant B is typically determined experimentally. Thus the physical equations describing the microphone are

$$R\frac{dQ}{dt} + \frac{Q}{B}h - E = 0,$$

$$m\frac{d^2h}{dt^2} = -\mu\frac{dh}{dt} - k(h-L) - \frac{Q^2}{2B} + F.$$

One choice of state variables is the charge on the capacitor and the displacement and velocity of the diaphragm so that $x_1 = Q$, $x_2 = h$, $x_3 = dh/dt$; the input variable u is the external acoustic force F. Thus the state equations are

$$\frac{dx_1}{dt} = -\frac{x_1 x_2}{RB} + \frac{E}{R},$$

$$\frac{dx_2}{dt} = x_3,$$

$$\frac{dx_3}{dt} = -\frac{\mu}{m}x_3 - \frac{k}{m}(x_2 - L) - \frac{1}{2}\frac{x_1^2}{Bm} + \frac{1}{m}u.$$

This is a third order nonlinear state model.

First, consider conditions for equilibrium. Suppose that $u(t) = 0$, $t \geqslant 0$.

Thus the microphone is in equilibrium if

$$\frac{dx_1}{dt} = -\frac{x_1 x_2}{RB} + \frac{E}{R} = 0,$$

$$\frac{dx_2}{dt} = x_3 = 0,$$

$$\frac{dx_3}{dt} = -\frac{\mu}{m} x_3 - \frac{k}{m}(x_2 - L) - \frac{1}{2}\frac{x_1^2}{Bm} = 0.$$

Clearly the conditions for an equilibrium state are that x_1 satisfy

$$x_1^3 - (2kBL)x_1 + 2kB^2 E = 0,$$

that x_2 satisfy

$$x_2 = L - \frac{mx_1^2}{2kB},$$

and that $x_3 = 0$. Since the algebraic equation for x_1 is a cubic there can be either one or three real solutions; hence there can be either one or three equilibrium states, depending on the parameter values.

Now suppose that $\bar{Q}, \bar{H}, 0$ denotes a fixed equilibrium state of the loudspeaker. Our objective is to obtain an approximate linear mathematical model which is valid so long as the actual state is near the equilibrium state. To this end define the variables z_1, z_2, z_3 from

$$x_1 = \bar{Q} + z_1,$$

$$x_2 = \bar{H} + z_2,$$

$$x_3 = 0 + z_3,$$

so that, in terms of these variational variables, the state equations are

$$\frac{d\bar{z}_1}{dt} = -\frac{(\bar{Q} + z_1)(\bar{H} + z_2)}{RB} + \frac{E}{R},$$

$$\frac{dz_2}{dt} = z_3,$$

$$\frac{dz_3}{dt} = -\frac{\mu}{m}(0 + z_3) - \frac{k}{m}(\bar{H} - L + z_2) - \frac{1}{2}\frac{(\bar{Q} + z_1)^2}{Bm} + \frac{1}{m}u.$$

Using the conditions for the equilibrium state obtain

$$\frac{dz_1}{dt} = -\frac{(\bar{Q}z_2 + \bar{H}z_1 + z_1 z_2)}{RB},$$

$$\frac{dz_2}{dt} = z_3,$$

$$\frac{dz_3}{dt} = -\frac{\mu}{m}z_3 - \frac{k}{m}z_2 - \frac{1}{2}\frac{(2\bar{Q}z_1 + z_1^2)}{Bm} + \frac{1}{m}u.$$

Now if z_1 and z_2 are significantly smaller than $\overline{Q}$ and $\overline{H}$ respectively then the nonlinear terms z_1z_2 and z_1^2 can legitimately be ignored to obtain the linearized state equations

$$\frac{dz_1}{dt} = -\left(\frac{\overline{H}}{RB}\right)z_1 - \left(\frac{\overline{Q}}{RB}\right)z_2,$$

$$\frac{dz_2}{dt} = z_3,$$

$$\frac{dz_3}{dt} = -\frac{\mu}{m}z_3 - \frac{k}{m}z_2 - \left(\frac{\overline{Q}}{Bm}\right)z_1 + \frac{1}{m}u.$$

Of course the coefficients in this linearized model depend on the particular equilibrium state chosen.

The response characteristics of the microphone are now analyzed using the above linearized state equations; use can be made of Laplace transforms in carrying out the analysis. In particular, the following transfer functions can be determined:

$$\frac{Z_1}{U} = -\frac{\overline{Q}/RBm}{d(s)},$$

$$\frac{Z_2}{U} = \frac{s + \overline{H}/RBm}{d(s)},$$

where the characteristic polynomial is

$$d(s) = \left[s^2 + \frac{\mu}{m}s + \frac{k}{m}\right]\left[s + \frac{\overline{H}}{RB}\right] - \frac{1}{Rm}\left(\frac{\overline{Q}}{B}\right)^2.$$

It is clear that proper operation of a microphone can occur only if the microphone is stable in some sense. Conditions for such stable operation, near the equilibrium state $\overline{Q}, \overline{H}, 0$ can be determined using the linearized state equations. In particular the Routh-Hurwitz criteria can be used to determine the conditions for stability. The analytical conditions are not developed here but it should be observed that if there is only one equilibrium state it is typically stable; if there are three equilibrium states then, typically, there is at least one unstable equilibrium state.

The electrical output of the microphone is the voltage across the capacitor. If this voltage is denoted by y then clearly

$$y = \frac{Q}{C} = \frac{Qz}{B}$$

or, in terms of the variational variables,

$$y = \left(\frac{\overline{Q}}{B}\right)z_2 + \left(\frac{\overline{H}}{B}\right)z_1,$$

where the small product term has been ignored.

Thus the transfer function from the force input to this voltage output is

$$\frac{Y}{U} = G(s) = \frac{(\overline{Q}/Bm)s}{d(s)},$$

where the previous expressions for X_1 and X_2 have been used. Now the frequency response function for the microphone is developed. The frequency response function is given by

$$G(j\omega) = \frac{(\overline{Q}/Bm)(j\omega)}{d(j\omega)}.$$

After some complex algebra the frequency response fucntion is determined as

$$G(j\omega) = Me^{j\phi},$$

where

$$M = \frac{(\overline{Q}/Bm)\omega}{D(\omega)}$$

and

$$D(\omega) = \left\{ \left[\frac{1}{R}\left(\frac{k}{m}\frac{\overline{H}}{B} - \frac{\overline{Q}^2}{B^2} \right) - \frac{1}{R}\left(\frac{\overline{H}}{B} + \frac{\mu}{m} \right)\omega^2 \right]^2 \right.$$
$$\left. + \omega^2 \left[\left(\frac{k}{m} + \frac{\mu}{m}\frac{\overline{H}}{RB} \right) - \omega^2 \right]^2 \right\}^{1/2}$$

The phase angle relation is not determined here. It is clear that the frequency response has the property that if $\omega = 0$ then $M = 0$; as $\omega \to \infty$ then $M \to 0$ and that M has a maximum value. A sample plot of the frequency response curve M versus ω is shown in Figure 7-4-2.

If the input is periodic so that it can be expressed in terms of a Fourier series of the form

$$u(t) = \sum_{i=0}^{\infty} F_i \cos(i\omega t + \xi_i)$$

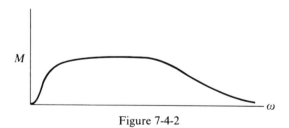

Figure 7-4-2

for constants F_i, ξ_i; then, assuming the microphone is operating near a stable equilibrium state, the steady state output response is

$$y(t) = \sum_{i=0}^{\infty} M_i F_i \cos(i\omega t + \phi_i + \xi_i),$$

where

$$M_i = M(i\omega),$$

$$\phi_i = \phi(i\omega).$$

Since a microphone is used to record acoustic waveforms for subsequent accurate reproduction it is desirable to have a microphone with a flat frequency response over the range of frequencies of importance.

Exercises

7-1. Consider the continuous flow stirred tank reactor described in Case Study 7-1. Assume the parameter values are

$$k = 0.2 \text{ liters}/(\text{mole-min}),$$

$$\frac{Q}{V} = 0.1 \text{ liters}/\text{min},$$

and suppose that the input concentrations are

$$C_A(t) = 0.5 \text{ moles}/\text{liter}, \quad C_B(t) = 0.0 \text{ moles}/\text{liter}$$

for $t \geqslant 0$.

(a) What are the equilibrium states of the reactor?
(b) What are linearized state equations which are valid when the reactor operates near the equilibrium state?
(c) Which equilibrium states are stable?
(d) If the initial concentrations in the reactor are

$$c_A(0) = 0.5 \text{ moles}/\text{liter},$$

$$c_B(0) = 0.5 \text{ moles}/\text{liter},$$

$$c_C(0) = 0.5 \text{ moles}/\text{liter},$$

what are the subsequent concentrations for $t \geqslant 0$?

7-2. Consider the vertical motion of a hot air balloon as developed in Case Study 7-2 with the parameter values

$$\tau = 1.0 \text{ hr}, \quad T_a = 60.0°\text{F}, \quad K = 0.05°\text{F}/\text{BTU},$$

$$W_a = 500.0 \text{ lb}, \quad W = 300.0 \text{ lb}, \quad \mu = 5.0 \text{ (lb-sec)}/\text{ft}.$$

(a) What are the conditions for the hot air balloon to be in equilibrium?

(b) Suppose that the heat transfer rate to the balloon from the heater is constant

$$Q(t) = 40.0 \text{ BTU/min} \qquad \text{for} \quad t \geqslant 0.$$

Describe the vertical motion of the balloon if initially

$$T(0) = 150.0°\text{F}, \quad h(0) = 500.0 \text{ ft}, \quad \frac{dh}{dt}(0) = 0.0 \text{ ft/sec}.$$

(c) Suppose that the heat transfer rate to the balloon is constant

$$Q(t) = 20.0 \text{ BTU/min} \qquad \text{for} \quad t \geqslant 0.$$

Describe the vertical motion of the balloon if initially

$$T(0) = 150.0°\text{F}, \quad h(0) = 500.0 \text{ ft}, \quad \frac{dh}{dt}(0) = 0.0 \text{ ft/sec}.$$

7-3. Consider the results of Case Study 7-3 as applied to the motion of a rocket about the Moon. On the surface of the Moon the acceleration of gravity is 5.3 ft/sec² and the radius of the Moon is 1080.0 miles.

(a) Assuming there is no thrust on the rocket, what is the required initial angular velocity for the rocket to move along a circular orbit at an altitude of 100.0 miles above the Moon's surface?

(b) Assuming there is no thrust on the rocket, suppose that the rocket is initially 100.0 miles above the surface of the Moon and moving so that initially

$$\frac{dr}{dt}(0) = 0.0 \text{ miles/sec}, \quad \frac{d\theta}{dt}(0) = 0.04 \text{ rad/min}.$$

Determine the motion of the rocket as it moves about the Moon; what is its altitude and angular orientation, as measured from some fixed reference, as a function of time? Consider a time period corresponding to one revolution of the rocket about the Moon.

7-4. Consider the model developed in Case Study 7-2 for the vertical motion of a hot air balloon. Develop state equations using as state variables the density of air inside the balloon, the altitude of the balloon above sea level, and the velocity of the balloon. Using these state equations determine conditions for equilibrium of the hot air balloon. Are there any obvious advantages in using this model compared to that given in Case Study 7-2?

7-5. Consider the model developed in Case Study 7-3 for the motion of a rocket. Develop state equations using as state variables the position and velocity variables

$$r, \quad \frac{dr}{dt}, \quad r\theta, \quad \frac{d(r\theta)}{dt}.$$

Using these state equations determine conditions for a circular orbit, assuming there is no thrust on the rocket. Are there any obvious advantages in using this model compared to that given in Case Study 7-3?

7-6. Consider the vertical ascent of a rocket from the Earth using the assumptions and notation of Case Study 7-3. Show that if the thrust on the rocket is always vertical, i.e., $\alpha(t) = 0$ for $t \geqslant 0$ and there is no tangential motion initially, i.e., $(d\theta/dt)(0) = 0$ then there is never any tangential motion, i.e., $d\theta(t)/dt = 0$ for $t \geqslant 0$, so that the vertical motion of the rocket is described by

$$m \frac{d^2 r}{dt^2} = -mg \left(\frac{R}{r} \right)^2 + T.$$

This is the model described in Exercise 6-9.

7-7. A uniform pendulum is attached via a pinned frictionless joint to a movable vehicle as shown

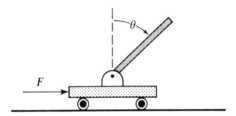

Here h is the displacement of the vehicle from a fixed reference position and θ is the angle of the pendulum from the vertical. Using Newton's law the motion of the pendulum and vehicle are described by

$$\frac{d}{dt} \left[m \left(R \cos \theta \frac{dh}{dt} + R^2 \frac{d\theta}{dt} \right) \right] = mgR \sin \theta,$$

$$\frac{d}{dt} \left[m \left(\frac{dh}{dt} + R \frac{d\theta}{dt} \cos \theta \right) + M \frac{dh}{dt} \right] = F,$$

where m is the mass of the pendulum, M is the mass of the vehicle, $2R$ is the length of the pendulum and F is the force applied to the vehicle. The force F is the input variable; the angle θ is the output variable.

(a) Choose state variables and write state equations.
(b) If the force $F = 0$, what are the equilibrium states?

    (c) Which equilibrium states are stable; which are unstable? What is the physical interpretation of your conclusions?

7-8. An electrostatic transducer can serve to convert electrical energy into acoustical energy. If Q is the electrical charge on a capacitor plate and h is the displacement of the movable element from some reference, then the transducer can be described by

$$L\frac{d^2Q}{dt^2} + R\frac{dQ}{dt} + \frac{Q}{B}h = V + e,$$

$$m\frac{d^2h}{dt^2} + \mu\frac{dh}{dt} + K(h - D) - \frac{Q^2}{2B} = 0.$$

The external voltage e is the input variable; the displacement h is the output variable. The other parameters, the inductance L, resistance R, reference position D, bias voltage V, diaphragm mass m, damping coefficient μ, diaphragm stiffness coefficient K, and scale factor B, are all assumed to be constant.

    (a) Choose state variables and write state equations.
    (b) If the input voltage $e(t) = 0$, for $t \geqslant 0$, what are the equilibrium states? Under what conditions on L, R, B, D, V, m, μ, K is there a single equilibrium state?
    (c) Assuming there is a single equilibrium state, as indicated in (b), obtain linearized state equations about that equilibrium state which could be used to analyze the zero input response of the transducer.

7-9. An asymmetrical satellite is moving in space without any external forces acting on it. Consider a coordinate system that is fixed to the satellite with its origin at the center of gravity of the satellite. Let I_1, I_2, I_3 denote the principal moments of inertia of the satellite about these three axes; let $\omega_1, \omega_2, \omega_3$ denote the angular velocities of the satellite about these three axes. There are on-board reaction wheel mechanisms which supply torques T_1, T_2, T_3 about these three axes, respectively. The torques are considered to be the input variables. The equations of motion of the satellite are the Euler equations

$$I_1\frac{d\omega_1}{dt} = (I_2 - I_3)\omega_2\omega_3 + T_1,$$

$$I_2\frac{d\omega_2}{dt} = (I_3 - I_1)\omega_3\omega_1 + T_2,$$

$$I_3\frac{d\omega_3}{dt} = (I_1 - I_2)\omega_1\omega_2 + T_3.$$

    (a) Write state equations for the satellite.
    (b) Assuming all torque inputs are zero, what are the equilibrium

states of the satellite? Interpret the physical meaning of equilibrium in this situation.

(c) Consider the equilibrium corresponding to a constant angular
velocity Ω_1 about the first principal axis and no rotation about
the other two principal axes. Develop linearized equations of
motion near this equilibrium condition.

(d) Develop linearized equations of motion about the equilibrium
condition corresponding to a constant angular velocity Ω_2 about
the second principal axis only.

(e) Develop linearized equations of motion about the equilibrium
condition corresponding to a constant angular velocity Ω_3 about
the third principal axis.

(f) Assuming that $I_1 < I_2 < I_3$, which equilibrium states of the satellite are stable? Use the linearized equations developed in (c), (d),
(e). What is the physical interpretation of your result?

7-10. A double pendulum is shown schematically below.

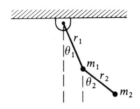

Masses m_1 and m_2 are connected by massless rods of length r_1 and r_2.
The equations of motion of the two masses, expressed in terms of the
angles θ_1 and θ_2 as indicated, are

$$(m_1 + m_2)r_1^2 \frac{d^2\theta_1}{dt^2} + m_2 r_1 r_2 \frac{d^2\theta_2}{dt^2} \cos(\theta_1 - \theta_2)$$

$$+ m_2 r_1 r_2 \left(\frac{d\theta_2}{dt}\right)^2 \sin(\theta_1 - \theta_2) = -(m_1 + m_2) g r_1 \sin\theta_1$$

$$m_2 r_2^2 \frac{d^2\theta_2}{dt^2} + m_2 r_1 r_2 \frac{d^2\theta_1}{dt^2} \cos(\theta_1 - \theta_2)$$

$$- m_2 r_1 r_2 \left(\frac{d\theta_1}{dt}\right)^2 \sin(\theta_1 - \theta_2) = -m_2 g r_2 \sin\theta_2.$$

(a) Choose state variables and write state equations.
(b) Show that $\theta_1 = 0$, $d\theta_1/dt = 0$, $\theta_2 = 0$, $d\theta_2/dt = 0$ defines an equilibrium state.
(c) Obtain linearized state equations which are valid if the pendulum
system is near its equilibrium state.

(d) What are the characteristic zeros of the double pendulum, assuming small oscillations of the masses? What is the physical interpretation of these characteristic zeros in terms of natural frequencies of the pendulum?

7-11. Another model for the vertical motion of a rocket is considered in this exercise. Here a constant gravitational field is considered but the change in mass of the rocket plus fuel is considered as fuel is burned. Let m denote the mass of the rocket, including fuel; let h denote the altitude of the rocket. Then the vertical motion of the rocket is described by

$$m\frac{d^2h}{dt^2} + c\frac{dm}{dt} = -mg,$$

where the exhaust velocity c and the acceleration of gravity g are

$$c = 1200.0 \text{ ft/sec}, \quad g = 32.0 \text{ ft/sec}^2.$$

The exhaust mass flow rate, $-dm/dt$, is the input variable; the altitude of the rocket is the output variable.

(a) Choose state variables and write state equations.
(b) Assume the weight of the rocket minus fuel is 50000.0 lb and the initial weight of the fuel is 100000.0 lb. The initial altitude and velocity of the rocket are $h(0) = 0$ and $(dh/dt)(0) = 0$. If the fuel flow rate is constant, i.e.,

$$\frac{dm}{dt}g = -8000.0 \text{ lb/sec}$$

as long as there is fuel available, what are the altitude and velocity of the rocket for $t \geq 0$? When is the fuel depleted? When does the rocket reach its maximum altitude? What is the maximum altitude of the rocket?

7-12. Another model for the vertical motion of a rocket is now considered. Here an inverse square law gravitational field is considered and the change in mass of the rocket plus fuel, as the fuel is burned, is also considered. Let m denote the mass of the rocket, including fuel; let h denote the altitude of the rocket. Then the vertical motion of the rocket is described by

$$m\frac{d^2h}{dt^2} + c\frac{dm}{dt} = -mg\left(\frac{R}{R+h}\right)^2,$$

where the exhaust velocity c, the acceleration of gravity constant g and the radius of the Earth R are

$$c = 1200.0 \text{ ft/sec}, \quad g = 32.0 \text{ ft/sec}, \quad R = 4000.0 \text{ miles}.$$

The exhaust mass flow rate, $-dm/dt$, is the input variable; the altitude of the rocket is the output variable.

(a) Choose state variables and write state equations.

(b) Assume the weight of the rocket minus fuel is 50000.0 lb and the initial weight of the fuel is 100000.0 lb. The initial altitude and velocity of the rocket are $h(0) = 0$ and $(dh/dt)(0) = 0$. If the fuel flow rate is constant, i.e.,

$$\frac{dm}{dt} g = -8000.0 \text{ lb/sec}$$

as long as there is fuel available, what are the altitude and velocity of the rocket for $t \geqslant 0$? When is the fuel depleted? When does the rocket reach its maximum altitude? What is the maximum altitude of the rocket?

7-13. Consider the irreversible chemical reaction of nitric oxide and oxygen

$$2 \text{ NO} + O_2 \rightarrow 2 \text{ NO}_2$$

which is assumed to occur in a closed vessel at constant temperature and pressure. If m_1, m_2, m_3 denote the molar concentrations of NO, O_2, NO_2, respectively, then the rate at which the chemical reaction proceeds is given by

$$\frac{dm_1}{dt} = -2k(m_1)^2 m_2,$$

$$\frac{dm_2}{dt} = -k(m_1)^2 m_2,$$

$$\frac{dm_3}{dt} = 2k(m_1)^2 m_2,$$

where the rate constant is

$$k = 2.5 \times 10^4 \text{ liter}^2/(\text{gm}^2\text{-mole}^2\text{-sec}).$$

(a) What are the equilibrium states of the reaction?

(b) If initially the molar concentrations in the reaction vessel are

$$m_1(0) = 0.01 \text{ moles/liter},$$

$$m_2(0) = 0.01 \text{ moles/liter},$$

$$m_3(0) = 0.0 \text{ moles/liter},$$

what are the molar concentrations in the reaction vessel for $t \geqslant 0$?

7-14. A projectile is fired at an angle $\theta = 45.0°$ from the horizontal, at the surface of the Earth. Let (x, y) denote the coordinates of the projectile, assuming a flat Earth. Consider a constant vertical gravitational force and a drag force which is proportional to the magnitude of the projectile's velocity and acts opposite to the direction of motion of the projectile. The equations of motion of the projectile are

$$\frac{d^2x}{dt^2} = -D\cos\alpha,$$

$$\frac{d^2y}{dt^2} = -D\sin\alpha - g,$$

where $g = 9.8$ meters/sec^2 is the acceleration of gravity and D is the magnitude of the drag induced deceleration given by

$$D = c\left[\left(\frac{dx}{dt}\right)^2 + \left(\frac{dy}{dt}\right)^2\right]^{1/2},$$

where the constant $c = 0.015$ 1/sec. The angle α is the angle between the velocity vector of the projectile and the horizontal; thus

$$\tan\alpha = \frac{dy/dt}{dx/dt}.$$

Assume the initial position and velocity of the projectile are

$$x(0) = 0, \quad \frac{dx}{dt}(0) = V\cos\theta,$$

$$y(0) = 0, \quad \frac{dy}{dt}(0) = V\sin\theta,$$

where $V = 250.0$ meters/sec is the initial velocity of the projectile.

(a) Choose state variables and write state equations.
(b) What is the maximum altitude of the projectile? How long does it take for the projectile to reach this altitude?
(c) What is the range of the projectile? How long does it take for the projectile to reach this range?
(d) Describe the path of the projectile.

7-15. In this exercise the two dimensional motion of a boat moving in a river of width L is considered. Assume there is a fixed coordinate system so that x and y denote the cross stream and down stream positions of the boat with respect to the coordinate system. It is assumed that there is a downstream river current with the velocity profile

$$W\sin\frac{\pi x}{L} \quad \text{for} \quad 0 \leqslant x \leqslant L.$$

Here W is the midstream velocity of the river. The equations of motion of the boat are

$$m\frac{d^2x}{dt^2} = T\cos\alpha - c\frac{dx}{dt},$$

$$m\frac{d^2y}{dt^2} = T\sin\alpha - c\left[\frac{dy}{dt} - W\sin\frac{\pi x}{L}\right].$$

Here m is the mass of the boat, T is the constant thrust of the boat's engine, α is the angle between the thrust vector and the cross stream coordinate axis and c is a drag coefficient for the boat. Assume the parameter values are

$$mg = 500.0 \text{ lb}, \quad T = 25.0 \text{ lb}, \quad L = 1.0 \text{ mile},$$
$$W = 2.0 \text{ mile/hr}, \quad c = 2.5 \text{ (lb-hr)/mile}.$$

Suppose that initially

$$x(0) = 0, \quad \frac{dx}{dt}(0) = 0, \quad y(0) = 0, \quad \frac{dy}{dt}(0) = 0.$$

The angle α is assumed to be the input variable.

(a) Choose state variables and write state equations.
(b) What is the path of the boat for a constant angle

$$\alpha(t) = 10.0° \qquad \text{for} \quad t \geqslant 0?$$

How long does it take for the boat to cross the river? What is the downstream location of the boat when the opposite shore is reached?

Chapter VIII

Other Differential State Models

In the previous chapters the state models have had a particular form, expressed in terms of first order ordinary differential equations. In those models the rate of change of each state variable could be expressed in terms of the values of the state and input variables at the same instant of time. However, there are physical features which cannot be captured using such models. In this chapter state models are briefly examined where the rate of change of each state variable is allowed to have a more general dependence; in particular the rate of change of each state variable may depend on the values of the state and input variables at the same and earlier instants of time.

No attempt is made to develop a theory of such differential state models. It suffices to mention that a general form for such models is given by

$$\frac{dx_1(t)}{dt} = f_1(x_1, \ldots, x_n, u)_{t'}$$

$$\vdots$$

$$\frac{dx_n(t)}{dt} = f_n(x_1, \ldots, x_n, u)_{t'}$$

$$y(t) = g(x_1, \ldots, x_n, u)_{t'}$$

where the above notation is meant to indicate the possibility that $f_1(x_1, \ldots, x_n, u)_t$ might depend on values of $x_1, \ldots, x_n, u$ at some time $t' < t$, etc. As usual u denotes the input variable and y denotes the output variable.

Depending on the particular specification the variables $x_1, \ldots, x_n$ (possibly together with additional information) define the state in the sense that the state summarizes the previous "history" of the system. Such nonclassical differential state models can be classified on the basis of the nature of the functions $f_1, \ldots, f_n, g$; for example, such models can be classified as linear or nonlinear state models.

It is not possible to develop a general theory for analyzing such models; but many of the systems concepts and analytical procedures discussed in the previous chapters may prove useful in special cases of the more general models considered in this chapter. For example, the notions of zero state response, zero input response, transient response, steady state response, equilibrium state, stability, oscillation, etc. are meaningful concepts. For linear state models certain transform methods might prove useful; for nonlinear state models it may be possible to develop and use linearized approximation models. Further, there are many special techniques that might prove useful for a limited class of special models with particular mathematical structures.

Several case study examples are now presented to indicate a few examples of the nonclassical differential state models. In each case the analysis is carried out using relatively simple procedures.

Case Study 8-1: Feedback Control of the Liquid Level in a Tank

In this example a modification is made to the example considered in Case Study 5-2. A vertical tank is considered, where liquid is allowed to drain through a hole in the bottom of the tank. In addition liquid flows into the tank as indicated in Figure 8-1-1. In this case assume that the liquid level is measured, e.g., using optical techniques or a pressure transducer, and the flow into the tank is automatically adjusted.

Let h denote the depth of liquid in the tank; thus, as discussed in Case Study 5-2, the flow relation is

$$A_s \frac{dh}{dt} = Q - A_h 0.6\sqrt{2gh} ,$$

where A_s and A_h are the cross sectional areas of the tank and hole respectively, and Q is the instantaneous flow rate into the tank. The flow

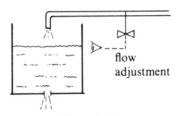

flow
adjustment

Figure 8-1-1

rate into the tank at time t is assumed to depend on the measured liquid level at the earlier time $t - T$ according to the relation

$$Q(t) = K\left[R - h(t - T)\right] + 0.6A_h\sqrt{2gR} \ .$$

Here $K > 0$ is a gain constant and $R > 0$ is a constant which represents the external input; R can be interpreted as the desired liquid level in the tank. The form of this dependence may seem somewhat arbitrary, but as will be demonstrated shortly, the liquid level in the tank does have certain desirable properties using this control algorithm. The flow at time t is assumed to depend on the measured level in the tank at the previous time $t - T$, since there is a time period required for the flow adjustment to actually reach the liquid in the tank. This time delay $T > 0$ is assumed constant. Our main interest in this example is to examine the effects of this time delay on the liquid level response properties.

The preceding two equations can be combined to yield

$$A_s \frac{dh(t)}{dt} = K\left[R - h(t - T)\right]$$

$$+ 0.6A_h\sqrt{2gR} - 0.6A_h\sqrt{2gh(t)} \ ;$$

the arguments of the dependent variable $h(t)$ are explicitly indicated. This is a first order nonlinear differential-difference equation. The state concept is first examined. Given a constant desired level R the actual liquid level in the tank can be determined for $t \geqslant 0$ only if the initial data $h(t)$, $-T \leqslant t \leqslant 0$ is specified; physically this corresponds to knowledge of the liquid level for a time period T previous to the initial time $t = 0$. It is not sufficient to specify only the value $h(0)$ in order to uniquely determine the future response. Thus the state at time 0 is an initial function $h(t)$ defined on the time period $-T \leqslant t \leqslant 0$; the state at time t' is the function $h(t)$ defined on $t' - T \leqslant t \leqslant t'$. This is the appropriate state concept in this example so that knowledge of the initial state and the input function uniquely determine the future states.

Some simple response properties are now determined. Suppose that initially $h(t) = R$ for $-T \leqslant t \leqslant 0$; then from the mathematical model it follows that $h(t) = R$ for all $t \geqslant 0$. Thus $h(t) = R$, $t \geqslant 0$ is a constant response, an equilibrium state. Note that in this case the liquid level in the tank exactly equals the desired level R considered as an external input function. To determine the response properties when $h(0) \neq R$ requires more detailed analysis of the mathematical model.

A linearization approach can be used here as in the previous chapters. In particular introduce the variable $z(t)$, $t \geqslant 0$ by

$$h(t) = R + z(t)$$

so that z denotes the variation of the liquid level from its equilibrium value.

Thus in terms of this variational variable

$$A_s \frac{dz(t)}{dt} = -Kz(t-T) + 0.6A_h\sqrt{2gR}$$

$$- 0.6A_h\sqrt{2g(R+z(t))} \ .$$

For sufficiently small values of z use the linear approximation

$$\sqrt{2g(R+z)} \simeq \sqrt{2gR} + 0.5\sqrt{\frac{2g}{R}}\ z.$$

Thus the linearized differential-difference equation

$$A_s \frac{dz(t)}{dt} = -KZ(t-T) - 0.3A_n\sqrt{\frac{2g}{R}}\ z(t)$$

results. This linearized model is linear in the dependent variable but it still involves the delayed term $z(t-T)$.

There are a number of ways to handle the delayed term in the analysis of the response properties. One procedure, which is often valid for small time delays is to approximate the delayed term as follows:

$$z(t-T) \simeq z(t) - \left(\frac{dz(t)}{dt}\right)T,$$

using the power series expansion formula for $z(t-T)$. Using this approximation the linearized model simplifies to

$$(A_s - KT)\frac{dz(t)}{dt} = -\left[K + 0.3A_h\sqrt{\frac{2g}{R}}\ \right]z(t).$$

Thus, if $T < A_s/K$ and $K > 0$ it follows that for any $z(0)$, $z(t) \to 0$ as $t \to \infty$; thus $h(t) \to R$ as $t \to \infty$. However, for larger delays satisfying $T > A_s/K$ the system is unstable. Such instability is a possible consequence of the presence of a time delay in a mathematical model. This approach to the analysis of models with time delays should be used with caution since it is restricted to small delays.

Another approach is now considered where the time delay need not be small. This approach generates the response on the successive time intervals $0 \le t < T$; then on $T \le t \le 2T$, etc. For example, suppose that the specified initial state is

$$h(t) = h_0(t), \qquad -T \le t \le 0,$$

so that

$$z(t) = h_0(t) - R, \qquad -T \le t \le 0.$$

Thus, for $0 \leqslant t < T$,

$$A_s \frac{dz(t)}{dt} = -0.3 A_h \sqrt{\frac{2g}{R}} \; z(t) - K\big[h_0(t - T) - R \big],$$

and the delayed term is now known; this linear equation is easily solved to obtain

$$z(t) = \exp\left[-0.3 \frac{A_h}{A_s} \sqrt{\frac{2g}{R}} \; t \right] \big(h_0(0) - R \big)$$

$$- \frac{K}{A_s} \int_0^t \exp\left[-0.3 \frac{A_h}{A_s} \sqrt{\frac{2g}{R}} \; (t - \lambda) \right] \big(h_0(\lambda - T) - R \big) \, d\lambda;$$

thus

$$h(t) = R + \exp\left[-0.3 \frac{A_h}{A_s} \sqrt{\frac{2g}{R}} \; t \right] \big(h_0(0) - R \big)$$

$$- \frac{K}{A_s} \int_0^t \exp\left[-0.3 \frac{A_h}{A_s} \sqrt{\frac{2g}{R}} \; (t - \lambda) \right] \big(h_0(\lambda - T) - R \big) \, d\lambda$$

which is the liquid level response for $0 \leqslant t \leqslant T$. In a similar fashion, for $T \leqslant t \leqslant 2T$,

$$A_s \frac{dz}{dt}(t) = -0.3 A_h \sqrt{\frac{2g}{R}} \; z(t) - K\big[h(t - T) - R \big],$$

where the delayed term is known from the previous computation. Thus the above equation can be integrated to obtain $z(t)$ and hence $h(t)$ for $T \leqslant t \leqslant 2T$. This method can be continued indefinitely to obtain the response on $2T \leqslant t \leqslant 3T$, etc. This method is conceptually attractive since it is not limited to assuming a small time delay; however, the approach does not lead to qualitative insight into the response characteristics due to the complex forms of the formulas.

This simple example does illustrate the way a time delay may be included in a mathematical model. Several solution techniques have also been indicated for such a mathematical model. It should be clear that mathematical models with time delays are considerably more difficult to analyze than if time delay were ignored. Elementary techniques have been used in the analysis here; a rather general theory of mathematical models containing time delays has been developed elsewhere in the mathematical literature.

Case Study 8-2: Automatic Regulation of Temperature in a Building

In Case Study 2-2 a mathematical model for the temperature in a building was developed. In this example that model is extended to include the fact that a furnace supplies heat to the building in order to regulate the inside temperature automatically. From the model developed in Case Study 2-2 the rate of change of the temperature inside the building is given by

$$\frac{dT}{dt} = -\frac{1}{\tau}[T - T_a] + KQ,$$

where T is the temperature inside the building, Q is the heat transfer rate to the building from the furnace; T_a is the constant temperature of the air surrounding the building, and K and τ are constants characterizing the thermal properties of the building. A realistic assumption is that the furnace operates in either an on-mode or an off-mode, depending on the measured temperature in the building. More specifically, the furnace is assumed to be on, consequently supplying a constant heat rate Q_0 to the building, if the temperature in the building is at least T_n degrees below the desired temperature T_d. The furnace is assumed to be off, consequently supplying no heat to the building, if the temperature in the building exceeds the desired temperature T_d by at least T_n degrees. If the temperature in the building satisfies $T_d - T_n \leqslant T \leqslant T_d + T_n$ the furnace is either off or on depending on whether the furnace was off or on a small instant previously; one way of specifying this relation mathematically is as follows.

$$Q = Q_0 \quad \text{if} \quad T \leqslant T_d - T_n$$
$$\text{or} \quad T_d - T_n < T < T_d + T_n \quad \text{and} \quad \frac{dT}{dt} > 0,$$

$$Q = 0 \quad \text{if} \quad T > T_d + T_n$$
$$\text{or} \quad T_d - T_n < T < T_d + T_n \quad \text{and} \quad \frac{dT}{dt} < 0.$$

Obviously the furnace is on in the first case and off in the second. This relationship is often characterized schematically by the hysteresis type curve shown in Figure 8-2-1. The maximum heat transfer rate Q_0, the desired temperature T_d and the hysteresis temperature T_n are all assumed to be constant. For reasons which will become clear later assume that the furnace heat transfer rate Q_0 is sufficiently large and the outside temperature T_a is sufficiently low so that the two inequalities

$$\tau K Q_0 > T_d + T_n - T_a,$$
$$T_a < T_d - T_n$$

are satisfied.

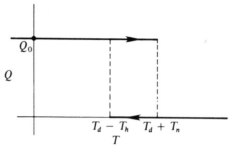

Figure 8-2-1

The mathematical model for the temperature in the building is thus given by

$$\frac{dT}{dt} = -\frac{1}{\tau}[T - T_a] + KQ_0$$

$$\text{if} \quad T \leqslant T_d - T_n,$$

$$\text{or} \quad T_d - T_n < T < T_d + T_n \quad \text{and} \quad \frac{dT}{dt} > 0,$$

$$\frac{dT}{dt} = -\frac{1}{\tau}[T - T_a]$$

$$\text{if} \quad T \geqslant T_d + T_n,$$

$$\text{or} \quad T_d - T_n < T < T_d + T_n \quad \text{and} \quad \frac{dT}{dt} > 0.$$

This is a nonlinear differential equation but it is certainly not in standard form since the expression for the rate of change dT/dt depends logically on T and dT/dt. In fact the state at time $t = 0$ consists not only of the temperature $T(0)$ but also the "state" of the furnace, i.e., whether it is initially on or off. Only if such an initial state is specified is it possible to determine the subsequent temperature in the building.

It is quite difficult to determine the general response properties, but one particular response can be easily determined. In particular suppose that initially the furnace is on and

$$T(0) = T_d - T_n.$$

If the furnace is initially on there is some time period $0 \leqslant t \leqslant t_1$ during which the furnace remains on so that $Q(t) = Q_0$, $0 \leqslant t \leqslant t_1$. The response during this time period is easily found by solving

$$\frac{dT}{dt} = -\frac{1}{\tau}[T - T_a] + KQ_0$$

so that for $0 \leqslant t \leqslant t_1$,

$$T(t) = (T_a + \tau KQ_0) + e^{-t/\tau}\big[(T_d + T_n) - (T_a + \tau KQ_0)\big].$$

This response function is valid so long as the furnace remains on, until the temperature in the building reaches $T_d + T_n$ degrees; thus the time t_1 at which the furnace turns off satisfies $T(t_1) = T_d + T_n$. Using the above expression for the building temperature this time can be determined as

$$t_1 = \ln\left[\frac{(T_a + \tau K Q_0) - (T_d - T_n)}{(T_a + \tau K Q_0) - (T_d + T_n)}\right].$$

In the above analysis the temperature in the building reaches $T_d + T_n$ degrees only if

$$\tau K Q_0 > T_d + T_n - T_a,$$

as assumed earlier. At time t_1 the furnace turns off and remains off for some time period $t_1 \leqslant t \leqslant t_2$ during which

$$\frac{dT}{dt} = -\frac{1}{\tau}[T - T_a].$$

This equation is easily solved with the condition that $T(t_1) = T_d + T_n$ to obtain, for $t_1 \leqslant t \leqslant t_2$,

$$T(t) = T_a + e^{-(t-t_1)/\tau}\left[(T_d + T_n) - T_a\right].$$

This response function is valid so long as the furnace remains off and the furnace remains off until the temperature in the building reaches $T_d - T_n$ degrees; thus the time t_2 at which the furnace turns on again satisfies $T(t_2) = T_d - T_n$. Using the above expression for the building temperature this time can be determined as

$$t_2 = t_1 + \tau \ln\left[\frac{(T_d + T_n) - T_a}{(T_d - T_n - T_a)}\right].$$

In the above analysis the temperature in the building reaches $T_d - T_n$ degrees only if

$$T_a < T_d - T_n,$$

as assumed earlier.

Now at time t_2 the furnace turns on again and $T(t_2) = T_d - T_n$. Notice that this is exactly the same as the initial state so that the response is necessarily repetitive or periodic. Thus $T(t) = T(t + t_2)$ for all $t \geqslant 0$ so that

$$t_2 = \tau \ln\left[\frac{(\tau K Q_0 + T_a - T_d + T_n)(T_d + T_n - T_a)}{(\tau K Q_0 + T_a - T_d - T_n)(T_d - T_n - T_a)}\right]$$

is the period. Thus the furnace is alternately on, then off, then on, etc. and the temperature in the building is maintained between $T_d - T_n$ and $T_d + T_n$. As T_n is decreased the building temperature is maintained in a narrow range but the period or cycle time of the furnace is reduced. The periodic temperature response is not sinusoidal but consists of two exponential segments as shown in Figure 8-2-2.

In this example a simple hysteresis type model of a furnace, operating in on-off mode, has been introduced. A simple technique has been demonstrated by which a certain periodic response was determined. Although the

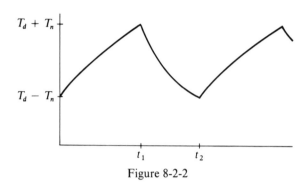

Figure 8-2-2

example considered here is quite specific it does illustrate the fact that nonlinear relations, such as the furnace model used here, may give rise to nonsinusoidal but periodic responses.

Case Study 8-3: Sampled Control of a Field Actuated DC Motor

In this example feedback control of the speed of a DC motor is studied, assuming that the armature current is constant. The notation and assumptions of Case Study 3-3 are used throughout. The basic idea is to control the speed of the motor at a desired value; this is achieved by measuring the actual motor speed, comparing it with the desired motor speed and generating a field voltage which is proportional to the error. It is assumed that a digital amplifier is used as the controller. The digital amplifier is assumed to operate as follows: the error signal is measured only at certain uniformly spaced times, these sampled values of the error are amplified and the resulting values are reconverted to a voltage applied to the field of the motor. This sampled data feedback system can be represented schematically by the block diagram in Figure 8-3-1.

Here u is the desired motor speed, y is the actual motor speed, $e = u - y$ is their difference, and v is the voltage applied to the field of the motor. The digital amplifier can be assumed to consist of an AC-DC converter, an amplification constant M, and a DC-AC converter.

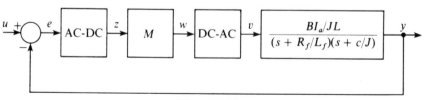

Figure 8-3-1

The AC-DC converter can be described as follows. The error signal $e(t)$, $t \geqslant 0$ is sampled at a uniform rate every T time units; T is the sample time, and sampling of the error occurs at the times

$$t_i = iT, \qquad i = 0, 1, 2, \ldots .$$

Assuming perfect sampling, then

$$z(t_i) = e(t_i), \qquad i = 0, 1, 2, \ldots .$$

The digital amplifier then processes the sampled error signal so that the output of the amplifier is proportional to the input to the amplifier; thus

$$w(t_i) = Mz(t_i), \qquad i = 0, 1, 2, \ldots .$$

Finally the DC-AC converter reconverts the discrete time signal $w(t_i)$ back to a continuous time voltage $v(t)$ by the zero order hold operation

$$v(t) = w(t_i) \qquad \text{if} \quad t_i \leqslant t \leqslant t_{i+1}, \quad i = 0, 1, 2, \ldots .$$

The applied field voltage $v(t)$, $t \geqslant 0$, is necessarily piecewise constant.

It is possible to show that this sampled data feedback system can be modelled using recursive equations defined at the sampling times t_i, $i = 0, 1, 2, \ldots .$ In order to obtain a recursive model recall that the differential equations which constitute a second order linear state model for a field actuated DC motor, as given in Case Study 3-3, are

$$\frac{dx_1}{dt} = -\frac{R_f}{L_f} x_1 + \frac{v}{L_f},$$

$$\frac{dx_2}{dt} = -\frac{c}{J} x_2 + \frac{BI_a}{J} x_1,$$

$$y = x_2,$$

where x_1 is the field current and x_2 is the motor speed. Now the equations describing the digital amplifier can be rewritten as

$$v(t) = M[u(t_i) - y(t_i)], \qquad t_i \leqslant t < t_{i+1}, \quad i = 0, 1, 2, \ldots .$$

Thus on a time period $t_i \leqslant t < t_{i+1}$, the field voltage $v(t) = v(t_i)$ is constant; hence on such a time period between sampling

$$\frac{dx_1}{dt} = -\frac{R_f}{L_f} x_1 + \frac{v(t_i)}{L_f},$$

$$\frac{dx_2}{dt} = -\frac{c}{J} x_2 + \frac{BI_a}{J} x_1.$$

It is easy to solve these differential equations for the responses over the time period $t_i \leqslant t < t_{i+1}$ in terms of the state at time t_i; the responses $x_1(t)$ and $x_2(t)$ clearly depend on $x_1(t_i)$, $x_2(t_i)$ and $v(t_i)$. The form of these responses for $t_i \leqslant t < t_{i+1}$ is given as follows:

$$x_1(t) = \phi_1(t - t_i, x_1(t_i), x_2(t_i), v(t_i))$$

$$= x_1(t_i)\exp\left(-\frac{R_f}{L_f}(t - t_i)\right) + \frac{v(t_i)}{R_f}\left[1 - \exp\left(-\frac{R_f}{L_f}(t - t_i)\right)\right],$$

$$x_2(t) = \phi_2(t - t_i, x_1(t_i), x_2(t_i), v(t_i))$$

$$= x_2(t_i)\exp\left(-\frac{c}{J}(t - t_i)\right)$$

$$+ x_1(t_i)\left(\frac{BI_a}{J}\right)\left(\frac{R_f}{L_f} - \frac{c}{J}\right)\left[\exp\left(-\frac{c}{J}(t - t_i)\right)\right.$$

$$\left. - \exp\left(-\frac{R_f}{L_f}(t - t_i)\right)\right]$$

$$+ v(t_i)\left\{\frac{BI_a}{cR_f} + \frac{BI_a}{(R_f/L_f - c/J)}\left[\frac{1}{JR_f}\left(\exp - \frac{R_f}{L_f}(t - t_i)\right)\right.\right.$$

$$\left.\left. - \frac{1}{cL_f}\exp\left(-\frac{c}{J}(t - t_i)\right)\right]\right\},$$

where the functions ϕ_1 and ϕ_2 are defined as indicated and it is assumed that

$$\frac{R_f}{L_f} \neq \frac{c}{J}.$$

Now substituting

$$v(t_i) = M[u(t_i) - x_2(t_i)]$$

into the above and taking the limit as $t \to t_{i+1}$ it follows that

$$x_1(t_{i+1}) = \phi_1[T, x_1(t_i), x_2(t_i), M(u(t_i) - x_2(t_i))],$$

$$x_2(t_{i+1}) = \phi_2[T, x_1(t_i), x_2(t_i), M(u(t_i) - x_2(t_i))]$$

for $i = 0, 1, 2, \ldots$.

In summary, the above equations are in the form of two first order linear recursive equations since each of $x_1(t_{i+1})$ and $x_2(t_{i+1})$ depend explicitly on $x_1(t_i)$, $x_2(t_i)$ and $u(t_i)$.

This recursive model could be analyzed in some detail using discrete transform techniques. However, it should be clear that the recursive equations are in a form which is directly useful, at least for numerical computation. Suppose that an input function $u(t)$, $t \geqslant 0$ is specified; then $u(t_i)$, $i = 0, 1, 2, \ldots$ is specified. For a given initial state $x_1(0), x_2(0)$ the state at time $t_1 = T$ can be directly determined from the recursive state model by proper evaluation of the two functions ϕ_1 and ϕ_2. Once $x_1(t_1), x_2(t_1)$ have been determined then the recursive state model can be used once again to determine the state at time $t_2 = 2T$. This procedure can be repeated iteratively to determine, at least in principle, $x_1(t_i), x_2(t_i)$ for all $i = 0, 1, 2, \ldots$. Note that the responses at the sampling times are determined solely from the recursive model; the responses between sampling times are also determined using the previous expressions

$$x_1(t) = \phi_1(t - t_i, x_1(t_i), x_2(t_i), M(u(t_i) - x_2(t_i))),$$
$$x_2(t) = \phi_2(t - t_i, x_1(t_i), x_2(t_i), M(u(t_i) - x_2(t_i))),$$

so long as $t_i \leqslant t < t_{i+1}, i = 0, 1, 2, \ldots$.

As one illustration of the response properties an equilibrium state is determined in the case where the input variable, the desired motor speed, is a constant:

$$u(t) = U, \qquad t \geqslant 0.$$

Thus at the sampling times $u(t_i) = U, i = 0, 1, 2, \ldots$. Now $\bar{x}_1, \bar{x}_2$ is an equilibrium state only if

$$x_1(t) = \bar{x}_1, \quad x_2(t) = \bar{x}_2, \qquad t \geqslant 0$$

is a response; thus necessarily at the sampling times

$$x_1(t_i) = \bar{x}_1, \quad x_2(t_i) = \bar{x}_2, \qquad i = 0, 1, 2, \ldots$$

From the discrete time model it follows that $\bar{x}_1, \bar{x}_2$ necessarily satisfy the two simultaneous algebraic equations

$$\bar{x}_1 = \phi_1(T, \bar{x}_1, \bar{x}_2, M(U - \bar{x}_2)),$$
$$\bar{x}_2 = \phi_2(T, \bar{x}_1, \bar{x}_2, M(U - \bar{x}_2)).$$

Of course, an alternative approach to determine the equilibrium state is to use the differential state model. If $\bar{x}_1, \bar{x}_2$ is an equilibrium state it follows that

$$0 = - \frac{R_f}{L_f} \bar{x}_1 + \frac{M}{L_f} (U - \bar{x}_2),$$

$$0 = - \frac{c}{J} \bar{x}_2 + \left(\frac{BI_a}{J} \right) \bar{x}_1.$$

These equations are easily solved to obtain

$$\bar{x}_1 = \frac{cMU}{cR_f + MBI_a},$$

$$\bar{x}_2 = \frac{MBI_a U}{cR_f + MBI_a}.$$

The usual purpose of the digital amplifier, used in the feedback structure under consideration, is to regulate the actual motor speed so that it is close to the desired motor speed, a constant in this case. The steady state error is given by

$$U - \bar{x}_2 = \frac{cR_f U}{cR_f + MBI_a} .$$

Hence, from the viewpoint of obtaining a small steady state error the value of the amplifier constant M should be chosen as large as possible. However, the transient response of the feedback system is also affected by the choice of M and should be considered in a detailed design.

The main purpose of this example has been to illustrate the effects of sampling, to point out how a recursive state model can be obtained for the responses at the sampling times, and to indicate an elementary approach to determining the response properties.

Exercises

8-1. Consider the model for the regulation of the liquid level in a tank as described in Case Study 8-1. Assume the following parameter values.

$$A_h = 1.0 \text{ in}^2, \quad A_s = 20.0 \text{ in}^2,$$

$$T = 1.5 \text{ sec}, \quad K = 6.0 \text{ in}^2/\text{sec}.$$

Assume that the desired liquid level in the tank is $R = 10.0$ in.

(a) What are the state equations?
(b) What are the linearized state equations if the liquid level is close to the equilibrium value?
(c) Is the regulated system stable?
(d) If the initial liquid level in the tank is 14.0 in, describe the time response of the liquid level in the tank for $t \geqslant 0$.

8-2. Consider the model developed in Case Study 8-2 for the automatic regulation of the temperature in a building. Assume the following parameter values.

$$\tau = 5.0 \text{ hr}, \quad T_a = 0.0°\text{F}, \quad K = 0.0005°\text{F}/\text{BTU},$$

$$Q_0 = 5200.0 \text{ BTU/hr}, \quad T_d = 70.0°\text{F}, \quad T_n = 5.0°\text{F}.$$

Suppose that the furnace is initially on and the initial temperature in the building is 65.0°F. How long is the furnace on; how long is the furnace off; what is the total cycle time of the furnace? Describe the time response for the temperature in the building for $t \geqslant 0$.

8-3. Consider the model developed in Case Study 8-3 for the sampled data control of a field actuated DC motor. Assume the following parameter values.

$$R_f = 20.0 \text{ ohms}, \quad L_f = 10.0 \text{ henry}, \quad I_a = 2.0 \text{ ohms},$$

$$c = 0.5 \text{ (ft-lb-sec)/rad}, \quad J = 4.0 \text{ (ft-lb-sec}^2)/\text{rad},$$

$$B = 1.0 \text{ (ft-lb)/amp}^2.$$

Suppose the amplifier constant is $M = 30.0$ (volt-sec)/rad and the sampling time is $T = 0.2$ sec. Describe, in as much detail as possible, the zero state response for a reference input function $u(t) = 5.0$ rad/-sec. What would be the effect of increasing the sampling time to $T = 0.5$ sec?

8-4. Two tanks, each of volume 50.0 ft³, are connected by flow lines as shown below.

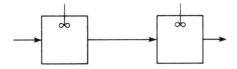

There is constant flow into and out of each tank of 5.0 ft³/min. Assume the tanks are completely filled with a salt solution which is kept at a uniform concentration in each tank by mixing. The pipe connecting the two tanks has a cross sectional area of 0.5 ft² and a length of 10.0 ft. The input is the concentration of salt in the inflow line; the output is the concentration of salt in the outflow line. Do not ignore the time it takes for the flow from the first tank to reach the second tank.

(a) Choose state variables and write state equations.
(b) What is the transfer function from the input to the output?
(c) If initially there is no salt in either tank and the concentration of salt in the inflow line is 1.0 lb/ft³ for $t \geqslant 0$, what is the concentration of salt in the outflow line for $t \geqslant 0$?

8-5. Consider the model for automatic regulation of temperature in a building as developed in Case Study 8-2. Consider the additional feature that the furnace does not actually turn on (or turn off) until a time period t_p has elapsed after the temperature reaches $T_d + T_n$ (or $T_d - T_n$). Such a model would take into account the time required for "start up" and "shut down" of the furnace.

(a) What is a state model for the temperature in the building with this assumption?
(b) Describe the periodic operation of the furnace. What is the period; how does the temperature in the building change with time?

8-6. Consider the electrical network connection indicated.

The input voltage is v_i and the output voltage is e_0. The AC-DC converter is described by the sampling equation

$$v_0(t_i) = v_i(t_i),$$

where $t_i = iT$ are the uniformly spaced sampling times with $T = 0.01$ sec, $i = 0, 1, \ldots$. The DC-AC converter is described by the zero order hold relation

$$e_i(t) = v_i(t_i) \qquad \text{for} \quad t_i \leqslant t < t_{i+1}.$$

The RC circuit is described in Case Study 2-3 with the parameter values $R = 10^4$ ohm and $C = 10^{-6}$ farad.

(a) Give recursive equations for the network.
(b) Suppose the input voltage $v_i(t) = 10.0$ volts for $t \geqslant 0$; describe conditions for equilibrium.
(c) Suppose the input voltage $v_i(t) = 10.0 \cos(100\pi t)$ volts for $t \geqslant 0$; describe the nature of the steady state output voltage.

Chapter IX

Discrete Time and State Models

All of the previous chapters have considered the development and use of state models for dynamic systems described in terms of differential equations. A number of case study examples have been presented which illustrate the general procedures. In this chapter several models are developed which illustrate the fact that, in certain cases, allowances for changes in the state of a system need be made only at discrete instants in time and that the state itself, as well as the input and output variables, may make discrete changes in their values at these times.

No attempt is made here to develop a general theory of discrete systems. Rather it suffices to mention that the general form for such state models is in terms of recursive equations, such as

$$x_1(t_{i+1}) = f_1(x_1(t_i), \ldots, x_n(t_i), u(t_i)),$$

$$\vdots$$

$$x_n(t_{i+1}) = f_n(x_1(t_i), \ldots, x_n(t_i), u(t_i)),$$
$$y(t_i) = g(x_1(t_i), \ldots, x_n(t_i), u(t_i)),$$

where u and y are the input and output variables, and $x_1, \ldots, x_n$ are the state variables for the system. All variables are defined at the discrete time instants $t_0, t_1, t_2, \ldots$. The variables may or may not be defined for other points in time; in any event the dynamic responses, as based on the above recursive equation model, do not depend on the variables except at the indicated discrete set of time instants. The values of u, y and $x_1, \ldots, x_n$ may be real or integer valued, or they may be restricted to a finite set, e.g., if they are logical variables.

No attempt is made to develop any theoretical results for such discrete models, but a few observations should be clear. Note that the state $x_1(t_k)$, $\ldots, x_n(t_k)$ at the instant t_k depends on the initial state $x_1(t_0), \ldots,$ $x_n(t_0)$ and the input values $u(t_0), \ldots, u(t_{k-1})$. Thus use of the state concept here is consistent with the previous usage in the sense that the state of a system summarizes the past "history" of the system. Discrete models can be classified as linear or nonlinear; for linear discrete models it is sometimes convenient to make use of a mathematical theory of transforms, the z transforms, to express the solution of a recursive equation explicitly in terms of the initial state and the input sequence. For nonlinear recursive equations this is usually not possible; in some cases linearization procedures may prove useful. It should be clear that most of the conceptual response properties mentioned in earlier chapters have an analog in discrete models. It is possible to define the notions of zero state response, zero input response, transient response, steady state response, general response, equilibrium state, stability, oscillation, etc. Notice that the form of the discrete model, as a recursive equation, is particularly convenient for the purpose of digital computer simulation.

Several case study examples are now presented to indicate a few of the fundamental types of discrete models. In each case the analysis is carried out in a rather ad hoc fashion without attempting to develop a general theory. The examples considered are sufficiently simple that powerful procedures are not required.

Case Study 9-1: Occupancy in a Hospital Unit

In this example a simple model for occupancy in a hospital unit is developed. Our interest is in determining how the occupancy of the hospital unit, i.e., the nonnegative integer number of beds occupied, depends on the number of discharges, emergency arrivals, scheduled arrivals, and the number of beds available in the hospital unit. It is natural to develop a discrete time model by considering the number of occupied beds at, for example, the end of each day. Thus let $x(t_i)$ denote the number of occupied beds at the end of the ith day. Suppose that $d(t_i), a(t_i), s(t_i)$ denote the number of discharges and deaths during the day, the number of emergency arrivals during the day and the number of scheduled arrivals during the day, respectively. Suppose that there are L total beds available in the hospital unit. Then it should be clear that the occupancy at the end of the day is the occupancy at the end of the previous day plus the emergency arrivals plus the scheduled arrivals minus the discharges, subject to the fact that there are only L beds. In mathematical terms it follows that

$$x(t_{i+1}) = \min\left[L, x(t_i) - d(t_i) + a(t_i) + s(t_i)\right].$$

Clearly, $x(t_i)$ is the state variable in the problem and the input can be considered to be the excess occupancy

$$u(t_i) = a(t_i) + s(t_i) - d(t_i)$$

so that

$$x(t_{i+1}) = \min\left[L, x(t_i) + u(t_i) \right]$$

which is in the standard form of a first order nonlinear discrete model. Clearly all variables are integer valued.

A few simple observations can be made. Suppose that $u(t_i) = 0$, $i = 0, 1, \ldots$; then any $x \leqslant L$ is an equilibrium state since if $x(t_0) \leqslant L$ then $x(t_i) = x(t_0)$, $i = 0, 1, \ldots$. Next suppose that $u(t_i) \geqslant 0$, $i = 0, 1, \ldots$; $x = L$ is an equilibrium state since if $x(t_0) = L$ then $x(t_i) = L$, $i = 0, 1, \ldots$.

As an example suppose that the hospital has $L = 100$ beds. Then if initially there are $x(t_0) = 90$ beds occupied and the input sequence of excess occupancy $u(t_i)$ for $i = 0, 1, \ldots, 5$ is

$$5, 8, -3, 5, -2, 3,$$

it is easily determined, using the state model, that the sequence of beds occupied $x(t_i)$ for $i = 0, 1, \ldots, 6$ is

$$90, 95, 100, 97, 100, 98, 100.$$

Notice also that on day number 2 there were 3 arrivals that could not be admitted, on day number 4 there were 2 arrivals that could not be admitted, on day number 6 there was 1 arrival that could not be admitted.

Much more detailed analysis could be carried out for various inputs $u(t_i)$. In fact this model can be extensively studied to determine the effect of the number of available beds L on the average occupancy of a hospital unit over a given period of time.

Case Study 9-2: An Inventory Model

A single commodity is stocked in a warehouse in order to satisfy a continuing demand. The daily demand for the commodity is the input u; the amount actually supplied is the output y. Assume that the replenishing of the stock in the warehouse takes place at the beginning of each day, so that the time t_i can be considered in daily increments.

The stock level x in the warehouse is now described, according to a commonly used inventory policy. The inventory policy depends on two fixed parameters s and $S > s$. The implementation of the inventory policy is as follows: if the available stock in the warehouse is in excess of s then no replenishment of stock is undertaken; if however the available stock is not greater than s then there is immediate procurement to bring the stock in the warehouse up to the level S.

Assume that $u(t_i)$, $y(t_i)$ and $x(t_i)$ are integer valued; $u(t_i)$ and $y(t_i)$ are necessarily nonnegative; x may be negative corresponding to a back-order which is immediately filled on restocking. Thus the change from the ith day to the $(i + 1)$st day is

$$x(t_{i+1}) = x(t_i) - u(t_i) \quad \text{if} \quad s < x(t_i),$$
$$x(t_{i+1}) = S - u(t_i) \quad \text{if} \quad x(t_i) \leqslant s.$$

Now the output which is the amount of stock supplied from the warehouse during the day is given by

$$y_s(t_i) = u(t_i) \quad \text{if} \quad s < x(t_i), \quad u(t_i) < x(t_i)$$
$$\text{or} \quad 0 < x(t_i) \leqslant s, \quad u(t_i) < S,$$
$$y_s(t_i) = x(t_i) \quad \text{if} \quad s < x(t_i), \quad u(t_i) \geqslant x(t_i),$$
$$y_s(t_i) = S \quad \text{if} \quad 0 < x(t_i) \leqslant s, \quad u(t_i) \geqslant S,$$
$$y_s(t_i) = u(t_i) - x(t_i) \quad \text{if} \quad x(t_i) \leqslant 0, \quad u(t_i) < S,$$
$$y_s(t_i) = S - x(t_i) \quad \text{if} \quad x(t_i) \leqslant 0, \quad u(t_i) \geqslant S.$$

Another variable of interest, the amount of stock added to the warehouse during the day, is

$$y_a(t_i) = 0 \quad \text{if} \quad s < x(t_i),$$
$$y_a(t_i) = S - x(t_i) \quad \text{if} \quad x(t_i) \leqslant S.$$

The above model is in the state variable form since $x(t_{i+1})$ depends on $x(t_i)$ and $u(t_i)$ while $y_a(t_i)$ and $y_s(t_i)$ depend on $x(t_i)$ and $u(t_i)$.

Several simple observations can be made about the state model. If $u(t_i) = 0$ for $i = 0, 1, \ldots$, then any $x(t_0) > s$ defines an equilibrium state since $x(t_i) = x(t_0)$, $i = 0, 1, \ldots$. Of course if the demand $u(t_i)$ is positive then there is no equilibrium state.

As an example suppose that the inventory policy is specified by the constants $s = 2$ and $S = 4$. Then if $x(0) = 3$ and the input sequence is $\{2, 1, 4, 2, 1, 0\}$ it is easily determined that the state values are $\{3, 1, 3, -1, 2, 3\}$ and the corresponding output values for y_s are $\{2, 1, 3, 3, 1, 0\}$ and for y_a are $\{0, 3, 0, 5, 2, 0\}$.

The simple inventory model can be used to determine the way in which demand can be met. There are clearly a number of modifications which could be made in the model to take into account such features as several commodities, different inventory policies, more realistic back order considerations, etc.

Case Study 9-3: Neuron Model

A basic component of the nervous system is the neuron, which transmits information in the form of electrical pulses from a dendrite to an axon. The mechanism by which these pulses are transmitted through the neuron is quite complex depending on the cellular chemistry of the neuron. Here a model of the neuron is based on the overall behavior of the neuron without looking at the detailed causal relationships. Typically a sequence of electrical pulses is transmitted to the neuron through the dendrite; only under certain conditions does the neuron transmit an electrical pulse along the axon, i.e., the neuron is said to fire. The basic assumption is that whether or not a neuron fires depends on the time since it last fired and the strength of the electrical pulses received by the neuron. Let the input u be the voltage level of the pulses received by the neuron; in this model attention is solely on whether the neuron fires or not so that the output variable y is 1 if the neuron fires and it is 0 otherwise. It is convenient to consider impulses as arriving at the neuron at discrete time instants $t_0, t_1, \ldots$ where $t_{i+1} - t_i = \tau$ is the synaptic delay time. As indicated earlier our basic assumption is that the firing of the neuron depends on the time since the neuron last fired; hence it is necessary to introduce the variable $x(t_i)$ which is the time since the neuron last fired. The neuron model is defined in terms of a threshold function $T(x)$ such that at time t_i the neuron fires if the received impulse at t_i exceeds the threshold function evaluated at the elapsed time since the neuron last fired; in mathematical terms this is expressed as

$$x(t_{i+1}) = \tau \qquad \text{if} \quad u(t_i) > T(x(t_i)),$$
$$x(t_{i+1}) = x(t_i) + \tau \qquad \text{if} \quad u(t_i) \leqslant T(x(t_i)).$$

Thus the expression for the output firing sequence is

$$y(t_i) = 1 \qquad \text{if} \quad u(t_i) > T(x(t_i)),$$
$$y(t_i) = 0 \qquad \text{if} \quad u(t_i) \leqslant T(x(t_i)).$$

A typical example of a threshold function is indicated in Figure 9-3-1.

It is clear that the model is in the state form; it is a first order state model. In this example the input is real valued, the output is binary valued and the state takes on values $\tau, 2\tau, 3\tau \ldots$.

There is no difficulty in analyzing this model in detail. As a simple illustration suppose that $u(t_i) > T(\tau)$, $i = 0, 1, \ldots$. Then $x = \tau$ is an equilibrium state since if $x(t_0) = \tau$ then $x(t_i) = \tau$, $i = 1, 2, \ldots$. That is, the neuron fires at each instant.

As another illustration suppose that the threshold function is given by

$$T(x) = \frac{E\tau}{x},$$

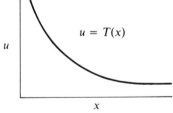

$$u = T(x)$$

Figure 9-3-1

where E is a constant, with units of voltage. Then the response if $x(t_0) = \tau$ and the input sequence $u(t_i)$ is $\{0.3E, 0.4E, 1.0E, 0.2E, 0.6E, 1.2E, \ldots\}$ can easily be determined. It can be verified that the corresponding state sequence $x(t_i)$ is $\{\tau, 2\tau, 3\tau, \tau, 2\tau, \tau, \ldots\}$ and the corresponding output firing pattern $y(t_i)$ is $\{0, 0, 1, 0, 1, 1, \ldots\}$.

This recursive state model is also useful for determining responses for more complicated threshold functions and more complicated input sequences. In addition the model can be extended, with some difficulty, to the important case of a network or interconnection of individual neurons.

Case Study 9-4: Coin Operated Dispenser

In this example consider a candy machine in which one can insert a nickel (N) or a dime (D). The price of the candy is 15¢. Thus, depending on how much money is inserted into the machine one obtains no return (0), a piece of candy (P), or a piece of candy (P) plus change (C). In this example the logic of this process is examined, the system is shown to be dynamic.

The input variable u is assumed to have the value D if a dime is inserted into the machine or N if a nickel is inserted into the machine. For convenience, two outputs are considered. The first output variable y_1 has the value 0 if no candy is delivered or the value P if candy is delivered. The second output variable y_2 has the value 0 if no change is delivered or the value C if change is delivered. This is clearly a discrete time process; it is convenient to consider only the time instants t_i for which the input variable is defined; that is, when some money is inserted into the machine. Clearly, our interest is not in the time itself but rather in the sequence of outputs as they depend logically on the input sequence.

The logic of the coin operated dispenser is as follows. It is clear that the outputs depend on the input but they also depend on how much money has been inserted into the machine previously. Hence it is important to consider the amount x of money accumulated in the coin dispenser; thus x can have the value 0¢, 5¢, or 10¢, and the following expressions clearly indicate how much money is accumulated in the machine, depending on how much was previously accumulated and how much was inserted into the machine.

$$
\begin{aligned}
x(t_{i+1}) = 0 \quad &\text{if} \quad x(t_i) = 10, \quad u(t_i) = D, \\
&\text{or} \quad x(t_i) = 10, \quad u(t_i) = N, \\
&\text{or} \quad x(t_i) = 5, \quad u(t_i) = D, \\
x(t_{i+1}) = 5 \quad &\text{if} \quad x(t_i) = 0, \quad u(t_i) = N, \\
x(t_{i+1}) = 10 \quad &\text{if} \quad x(t_i) = 5, \quad u(t_i) = N, \\
&\text{or} \quad x(t_i) = 0, \quad u(t_i) = D.
\end{aligned}
$$

Clearly the above constitute the state equations for this system since the outputs can be expressed in terms of $x(t_i)$ and $u(t_i)$, namely

$$y_1(t_i) = P \quad \text{if} \quad x(t_i) = 5, \quad u(t_i) = D,$$
$$\text{or} \quad x(t_i) = 10, \quad u(t_i) = N,$$
$$\text{or} \quad x(t_i) = 10, \quad u(t_i) = D,$$
$$y_1(t_i) = 0 \quad \text{otherwise};$$

and

$$y_2(t_i) = C \quad \text{if} \quad x(t_i) = 10, \quad u(t_i) = D,$$
$$y_2(t_i) = 0 \quad \text{otherwise}.$$

Note that the state $x(t_{i+1})$ depends on $x(t_i)$ and $u(t_i)$ and the outputs $y_1(t_i)$ and $y_2(t_i)$ each depend on $x(t_i)$ and $u(t_i)$. Thus the above constitutes a first order state model for the coin operated dispenser.

Since all of the variables in the problem can take values only in a finite set such a model is said to be a digital model or a sequential machine model.

Analysis of the response properties of the model is straightforward. As an illustration consider the response assuming that $x(t_0) = 0$ and that the input sequence $u(t_i)$ is given by $\{N, N, D, D, N, D, \ldots\}$. Then it is relatively easy to show that the resulting state sequence $x(t_i)$ is $\{0, 5, 10, 0, 10, 0, 10, \ldots\}$ and the output sequences for $y_1(t_i)$ and $y_2(t_i)$ are $\{0, 0, 0, P, 0, P, 0, \ldots\}$ and $\{0, 0, 0, C, 0, 0, 0, \ldots\}$, respectively.

This example is a very simple illustration of a large class of dynamic systems which can be described completely in terms of elementary logical expressions. Use of the formal mathematical modeling process, identifying the input, output and state, has been demonstrated.

Case Study 9-5: A Binary Communications Channel

There are many ways of modeling communications channels; in this case study the message to be transmitted over the communications channel is a sequence defined on the binary alphabet of symbols 0 and 1. This message is first encoded according to a specified encoder algorithm; the result is then transmitted over the communication channel and decoded using some decoder algorithm to obtain the received message. A very simple form for the encoder and decoder algorithms is assumed. The input is the message to be transmitted; the output is the received message. The objective is to characterize how the received message depends on the transmitted message and the channel characteristics. Under ideal circumstances the received message is exactly the transmitted message. However, our model of the communication channel allows for imperfect transmission. A schematic representation of the communications process is indicated in Figure 9-5-1.

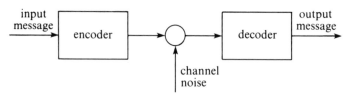

Figure 9-5-1

For simplicity a discrete time base defined by the nonnegative integers $i = 0, 1, 2, \ldots$ is used. Thus the input or transmitted message is a sequence $u(0), u(1), \ldots, u(i), \ldots$ and the output or received message is a sequence $y(0), y(1), \ldots, y(i), \ldots$. The communication channel is now described. First, the message to be transmitted is encoded as follows. The output of the encoder at time i is the (modulo 2) sum of the input message at time i and the input message at time $i - 1$. If $w(i)$ denotes the output of the encoder at time i then

$$w(i) = u(i) + u(i - 1).$$

The data sequence $w(0), w(1), \ldots, w(i), \ldots$ is transmitted and received as the input to the decoder in a possibly corrupted form; in particular if $z(i)$ denotes the received input to the decoder at time i then

$$z(i) = w(i) + v(i),$$

where $v(0), v(1), \ldots, v(i), \ldots$ represents the transmission noise process, i.e., $v(i) = 0$ if no transmission error is made while $v(i) = 1$ if a transmission error is made. Thus the input to the decoder is the (modulo 2) sum of the encoder output and a noise sequence. Finally, the output of the decoder at time i, which determines the output or received message, is the (modulo 2) sum of the decoder input at time i and the decoder output at time $i - 1$; that is,

$$y(i) = z(i) + y(i - 1).$$

The above algebraic expressions, written in terms of modulo 2 arithmetic, are taken as the model of the communication channel. The model may be written in state variable form by defining the state variables at time i as

$$x_1(i) = w(i) + u(i), \quad x_2(i) = y(i) + z(i).$$

Thus, the following (modulo 2) arithmetic calculations can be made:

$$\begin{aligned}
x_1(i + 1) &= w(i + 1) + u(i + 1) \\
&= u(i + 1) + u(i) + u(i + 1) \\
&= u(i)
\end{aligned}$$

and

$$x_2(i + 1) = y(i + 1) + z(i + 1)$$
$$= z(i + 1) + y(i) + z(i + 1)$$
$$= y(i)$$
$$= x_2(i) + z(i)$$
$$= x_2(i) + w(i) + v(i)$$
$$= x_2(i) + x_1(i) + v(i) + u(i).$$

Thus the state model is given by the second order recursive equations

$$x_1(i + 1) = u(i),$$
$$x_2(i + 1) = x_1(i) + x_2(i) + v(i) + u(i),$$
$$y(i) = x_1(i) + x_2(i) + v(i) + w(i).$$

In this formulation the noise process $v(0), v(1), \ldots, v(i) \ldots$ is assumed to be given as part of the specification of the communication channel. Thus if the output message sequence $u(0), u(1), \ldots u(i), \ldots$ and the initial state $x_1(0)$ and $x_2(0)$ are specified it is clear that the future state response and output message sequence are uniquely determined from the above state equations.

It is clear that the performance of this particular communication channel depends significantly on the characteristics of the noise process. First, the noise free case is examined where $v(i) = 0, i = 0, 1, \ldots$. If $x_1(i) = x_2(i)$ then necessarily $y(i) = u(i)$ and transmission is perfect. Thus to show that the received message output is identical to the transmitted message input it suffices to show that $x_1(i) = x_2(i)$ for all $i = 0, 1, \ldots$. Suppose that the encoder and decoder are initialized so that

$$x_1(0) = 0, \quad x_2(0) = 0.$$

Thus, from the state equations it follows that

$$x_1(1) = u(0) = x_2(1).$$

More generally, if $x_1(i) = x_2(i)$ then, from the state equations,

$$x_1(i + 1) = u(i) = x_2(i + 1).$$

Hence, in the noise free case, if the encoder and decoder are properly initialized so that $x_1(0) = 0$ and $x_2(0) = 0$ the output or received message is identical to the input or transmitted message.

In the case where there is transmission noise the output or received message is not the same as the input or transmitted message. As an example, a particular noise process is assumed, namely the noise sequence

$$v(i) = 1, \qquad i = 0, 1, \dots$$

corresponding to an error in each transmission. Consider the particular input message

$$1\ 0\ 1\ 1\ 1\ 0\ \dots\ .$$

The resulting output message can be determined by using the state equations. If the initial state is $x_1(0) = 0$ and $x_2(0) = 0$ then the output message can be determined to be the sequence

$$0\ 0\ 0\ 1\ 0\ 0\ \dots\ .$$

There are three errors in the transmitted message even though there are six noise induced errors.

It is not suggested that the encoder and decoder algorithms which have been considered are reasonable schemes; in fact they are too simple for accurate transmission of message data with substantial channel noise. Nevertheless, the intention has been to demonstrate that a simple binary communications channel can be modeled as a dynamical system and the state concept can be profitably used.

Case Study 9-6: Discrete Population Model

In Case Study 5-1 a population model was developed by assuming that the population was real valued and changed continuously in time. In this example a completely different viewpoint is taken. It is assumed that the population is integer valued and that changes in the population size occur at discrete instants of time; these changes occur either as a consequence of a birth event or a death event.

Let P denote the integer population; our interest is in determining how the population changes in time. The following assumptions about these changes are made. If a birth occurs at time t^B then the next birth is assumed to occur a time T^B later, where this time between births depends on the population $P(t^B)$, i.e., the next birth occurs at time $t^B + T^B(P(t^B))$. Similarly, if a death occurs at time t^D then the next death occurs at time $t^D + T^D(P(t^D))$, i.e., the time T^D between deaths depends on the population $P(t^D)$. The two functions T^B and T^D are important in characterizing the changes in the population. Now obviously if a birth occurs the population is incremented by one; if a death occurs the population is decremented by one. It is convenient to focus on the time instants $t_1, t_2, \dots$, where either a birth or a death occur. Unfortunately, it is not known a priori what these times are since they themselves depend on the changing population.

To derive a mathematical model for the population changes it should be clear that it is important to know the current population as well as the times when the next birth and death occur. Thus a mathematical description of the model is somewhat more complicated than might be at first apparent. It should be clear that

$$P(t_{i+1}) = P(t_i) + 1$$

if a birth occurs at t_{i+1},

$$P(t_{i+1}) = P(t_i) - 1$$

if a death occurs at t_{i+1}; and

$$t_{i+1} = t^B + T^B(P(t^B))$$

if a birth occurs at t_{i+1},

$$t_{i+1} = t^D + T^D(P(t^D))$$

if a death occurs at t_{i+1}. But note that whether a birth or a death occurs at t_{i+1} depends on the times and populations at the most recent birth and death events. Hence, to completely specify the model there is a need to define t_{i+1} in terms of whether a birth or death occurs; this consideration is avoided here since it merely complicates the model and in specific cases it is clear from the context if t_{i+1} is a time of birth or death. Our intention here is not to develop a complete state model but rather to indicate that knowledge of the current population only, under the indicated assumptions, is not sufficient to determine the future population. In particular, the state at time t consists of the population at time t and the times at which the next birth and death are scheduled. Only if such initial state information is specified is the future population uniquely determined. Since events, i.e., changes in the state, occur at discrete but a priori unknown time instants this kind of model is said to be a discrete event model. Such models typically involve more complex logic than the models considered previously.

It would be possible to make some general observations about the response properties based on the above model, but it is probably more helpful to carry out a simple simulation.

As an illustration suppose that T^B and T^D are assumed to be inversely proportional to the population P and that there are 50 births per hundred individuals per year and 20 deaths per hundred individuals per year. Thus $T^B(P) = 2.0/P$ and $T^D(P) = 5.0/P$. Assuming that the initial time is taken to be $t_0 = 0$ and the initial population is $P(0) = 100$, the calculations indicating the changes in the population are best demonstrated by considering an events list. The calculations should be self-explanatory; the results are indicated in Figure 9-6-1. A plot of the resulting response is indicated in Figure 9-6-2.

As expected, the population is, on the average, tending to increase since the average birth rate exceeds the average death rate.

Of course, additional complexities could be added to this population model to take into account sex differences, gestation times, age structures, etc. Such population models are extremely important in demographic studies.

Time of Occurrence of Event	Type of Event	Population Size	New Event Scheduled	Future Events List
0	Start	100	B @ 0.02 D @ 0.05	B @ 0.02 D @ 0.05
0.02	Birth	101	B @ 0.0399	B @ 0.0399 D @ 0.05
0.0399	Birth	102	B @ 0.0597	D @ 0.05 B @ 0.0597
0.05	Death	101	D @ 0.0998	B @ 0.0597 D @ 0.0998
0.0597	Birth	102	B @ 0.0795	B @ 0.0795 D @ 0.0998
0.0795	Birth	103	B @ 0.0992	B @ 0.9992 D @ 0.0998
0.0992	Birth	104	B @ 0.1188	D @ 0.0998 B @ 0.1188
0.0998	Death	103	—	—

Note: Information in the columns for Population Size and Future Events List, at a given time, constitute the state of the process at that time.

Figure 9-6-1

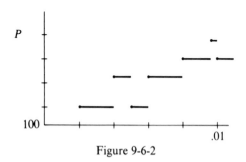

Figure 9-6-2

Exercises

9-1. A model for occupancy in a hospital unit was developed in Case Study 9-1. Suppose that the number of hospital beds is $L = 100$. If initially there are 95 occupied beds and the excess occupancy $u(t_i)$ for $i = 0, 1,$

2, . . . , 8 is given by the sequence

$$1, 4, -1, 6, -3, 2, -2, -4, 0$$

determine the resulting number of occupied beds on the succeeding days. During this time period how many emergency or scheduled arrivals cannot be admitted to the hospital?

9-2. Consider the inventory model as described in Case Study 9-2, where the inventory policy is defined by the policy parameters

$$s = 3, \quad S = 6.$$

Assume that the daily demand $u(t_i)$ for $i = 0, \ldots, 7$ is given by the sequence

$$2, 1, 3, 0, 4, 2, 5, 2.$$

Suppose that the initial stock level is 4 units. Determine the stock level and the number of units actually supplied on a daily basis for the succeeding days. During this time period how many total units were backordered, i.e., not immediately filled on demand?

9-3. Consider the binary communications channel as described in Case Study 9-5. Suppose that the noise process is given by the periodic sequence

$$0\ 0\ 0\ 1\ 1\ 0\ 0\ 0\ 1\ 1\ldots .$$

If the initial encoder and decoder states are

$$x_1(0) = 0, \quad x_2(0) = 0$$

and the input message is the periodic data sequence

$$1\ 1\ 1\ 0\ 0\ 1\ 1\ 1\ 0\ 0\ldots$$

determine the succeeding states of the encoder and decoder and the output message. What is the average number of errors in the output message for this particular noise process and this particular input message?

9-4. A model for population change is described in Case Study 9-6. Suppose the time between births and the time between deaths are given by

$$T^B(P) = \frac{5.0}{P},$$

$$T^D(P) = \frac{2.0}{P}.$$

Assuming the initial population consists of 100 individuals determine the resulting population level for a time period of 0.01 year.

9-5. Develop a state model for a coin operated dispenser, as considered in
Case Study 9-4 assuming that the machine accepts a nickel (N), a
dime (D) or a quarter (Q). The price of candy is 15¢. Determine the
zero state response assuming that the input $u(t_i)$ for $i = 0, 1, \ldots, 6$ is
given by the sequence

$$N, N, Q, D, N, D, Q.$$

Appendix A

Units, Dimensions, and Conversion Factors

In the development of mathematical models to describe physical processes one arrives at a set of mathematical equations which relate the several variables required in describing the process. These equations imply that the values of the several variables are such that the mathematical equations should be numerically satisfied. But it is also required that the various terms in the equations should be consistent in terms of the dimensions and units.

Dimensions are the basic concepts of measurement such as length or distance, time, mass, force, etc. Units are the means of expressing the dimensions, such as feet or meters for length, seconds or hours for time, etc. The following brief description should be helpful, but it is not intended to be complete.

The simple equation $A = B$ means that the numerical values of A and B are the same, and also the units in which those numerical values are expressed are equal. The recognition of this fact is sometimes helpful in focusing on obvious errors in a mathematical development; if two terms are added in a mathematical equation then those two terms must have the same dimensions; if they do not then the equation cannot be correct. In the presentation in the previous chapters there was no general concern with the units used to define the various variables, but it is important to be careful with the units in particular calculations. Thus the following conversion factors and physical constants should be of value.

Length:
    1 foot = 0.30048 meter
    1 mile = 5280 feet = 1.609 km

Area:

$1 \ m^2 = 10.76 \ ft^2$

Volume:

$1 \ liter = 10^{-3}m^3 = 0.0351 \ ft^3 = 61 \ in^3$

Velocity:

$1 \ meter/sec = 3.281 \ ft/sec$

$1 \ mile/min = 60 \ mile/hr. = 88 \ ft/sec$

Acceleration:

$1 \ meter/sec^2 = 3.281 \ ft/sec^2$

Force:

$1 \ newton = 0.2247 \ lb. = 1 \ kg\text{-}m/sec^2$

$1 \ lb = 4.45 \ newton = 1 \ slug\text{-}ft/sec^2$

Mass:

$1 \ kg = 0.0685 \ slug$

Voltage:

$1 \ volt = 1 \ amp\text{-}ohm = 1 \ coulomb/farad$

Current:

$1 \ ampere = 1 \ coulomb/sec = 1 \ volt/ohm = 1 \ volt\text{-}sec/henry$

Temperature (Change):

$1°C = 1.8°F$

$1°K = 1.8°R$

Energy:

$1 \ joule = 1 \ newton\text{-}meter = 0.72 \ ft\text{-}lb$

$= 0.239 \ calories = 9.25 \times 10^{-4} \ BTU$

Power:

$1 \ watt = 1 \ joule/sec = 1 \ volt\text{-}amp$

$= 0.72 \ ft\text{-}lb/sec = 0.139 \times 10^{-2} \ horsepower$

Atmospheric Pressure:

$1.013 \times 10^5 \ newtons/m^2 = 14.7 \ lb./in^2$

Absolute Zero of Temperature:

$-273.2°C = -460°F = 0°K = 0°R$

Acceleration Due to Gravity (Sea Level):

$9.78 \ m/sec^2 = 32.2 \ ft/sec^2$

Mean Radius of Earth:

$6.37 \times 10^6 \ meter = 3959 \ miles$

Density of Water:

$62.4 \ lb/ft^3$

Appendix B

Laplace Transforms

An extremely convenient tool for analysis of the response characteristics of linear time invariant differential equations is the Laplace transformation. This tool is used throughout the text; here a brief review of the important features of the Laplace transform used in the text is given. The treatment of Laplace transforms presented here is not complete; thorough treatment of the subject is available in the book by Kaplan.

The Laplace transformation converts or transforms a real function of a real variable into a complex function of a complex variable. In particular let $f(t)$ be a real function defined for $t \geq 0$; then the Laplace transform of this function is defined by

$$\mathcal{L}[f(t)] = \int_0^\infty f(t)e^{-st}\,dt$$

and is often written as $\mathcal{L}[f(t)] = F(s)$. This integral is defined for complex values of s so that $F(s)$ is a complex function and the methods of complex analysis are applicable. The Laplace transform is defined for a wide class of functions, so long as the complex values s are restricted to a region in the complex plane for which the indefinite integral in the definition converges.

Several important properties of the Laplace transform are now indicated.

1. Let a be an arbitrary real number and let $f(t)$, $t \geq 0$, have a Laplace transform. Then

$$\mathcal{L}[af(t)] = a\mathcal{L}[f(t)].$$

2. Let $f(t)$ and $g(t)$ be defined for $t \geq 0$, each having a Laplace transform. Then

$$\mathcal{L}[f(t) + g(t)] = \mathcal{L}[f(t)] + \mathcal{L}[g(t)].$$

The properties 1 and 2 above guarantee that the Laplace operator is a linear operator; linearity is a critical property of the Laplace transform.

3. Let a be an arbitrary real number and let $f(t)$, $t \geqslant 0$, have a Laplace transform $F(s)$. Then

$$\mathcal{L}\left[e^{-at}f(t)\right] = F(s + a).$$

4. Suppose that $f(t)$, $t \geqslant 0$, has a derivative defined for all $t \geqslant 0$; if the derivative has a Laplace transform then

$$\mathcal{L}\left[\frac{d}{dt}f(t)\right] = s\mathcal{L}\left[f(t)\right] - f(0).$$

5. Suppose that $f(t)$, $t \geqslant 0$, has an nth derivative defined for all $t \geqslant 0$; if the nth derivative has a Laplace transform then

$$\mathcal{L}\left[\frac{d^n}{dt^n}f(t)\right] = s^n\mathcal{L}\left[f(t)\right] - s^{n-1}f(0) - \cdots - \frac{d^{n-1}}{dt^{n-1}}f(0),$$

where $(d^i/dt^i)f(0)$ denotes the ith derivative of $f(t)$ evaluated at $t = 0$.

6. Let $h(t) = 1$, if $t \geqslant 0$ and $h(t) = 0$, if $t < 0$, denote the unit step function. Suppose that $a > 0$ and that $f(t)$, $t \geqslant 0$, has a Laplace transform. Then the delayed function $f(t - a)h(t - a)$ has the Laplace transform

$$\mathcal{L}\left[f(t - a)h(t - a)\right] = e^{-sa}\mathcal{L}\left[f(t)\right].$$

7. Let $f(t)$, $t \geqslant 0$, have the Laplace transform $F(s)$ and let $g(t)$, $t \geqslant 0$, have the Laplace transform $G(s)$. Then the convolution integral

$$\int_0^t f(\tau)g(t - \tau)\,d\tau$$

has the Laplace transform

$$\mathcal{L}\left[\int_0^t f(\tau)g(t - \tau)\,d\tau\right] = F(s)G(s).$$

8. Suppose that $f(t)$, $t \geqslant 0$, has a Laplace transform $F(s)$ and that the limit

$$\lim_{t \to \infty} f(t)$$

exists. Then the final value theorem

$$\lim_{t \to \infty} f(t) = \lim_{s \to 0} sF(s)$$

holds.

The above properties are often useful in developing the Laplace transform of a given function. However, the inversion problem is also of substantial interest and importance. Given the complex function $F(s)$ (which is a Laplace transform function) it is desired to find a function $f(t)$ defined for $t \geqslant 0$ such that

$$\mathcal{L}\left[f(t)\right] = F(s).$$

Sometimes this inversion property is written as

$$f(t) = \mathcal{L}^{-1}[F(s)].$$

There are general inversion procedures which use the methods of complex analysis; these procedures are discussed in the book by Kaplan. Here it suffices to mention that tables of Laplace transform pairs, such as in Figure B-1, are often useful in conjunction with the properties mentioned earlier. In addition, in many areas of application, it is important to be able to determine the inverse Laplace transform in the case where $F(s)$ is the ratio of two real polynomial functions in s of the form

$$F(s) = \frac{b_m s^m + b_{m-1} s^{m-1} + \cdots + b_0}{s^n + a_{n-1} s^{n-1} + \cdots + a_0} = \frac{N(s)}{D(s)},$$

where $m < n$ and $b_m \neq 0$. For such complex function it follows that there are m complex values $z_1, \ldots, z_m$ where $N(z_i) = 0$, $i = 1, \ldots, m$; these are the zeros of $F(s)$. Also, there are n complex values $p_1, \ldots, p_n$ where $D(p_i) = 0$, $i = 1, \ldots, n$; these are the poles of $F(s)$. Assuming that the polynomials $N(s)$ and $D(s)$ have real coefficients it follows that the poles and zeros consist of real values and values which occur as complex conjugate pairs. In any event, the polynomials $N(s)$ and $D(s)$ can be written in factored form so that

$$F(s) = \frac{b_m(s - z_1)(s - z_2) \cdots (s - z_m)}{(s - p_1)(s - p_2) \cdots (s - p_n)}.$$

Now suppose that the complex poles $p_1, \ldots, p_n$ are *distinct*. Then $F(s)$ can be written in the partial fraction expansion

$$F(s) = \frac{A_1}{s - p_1} + \cdots + \frac{A_n}{s - p_n},$$

where the residue constants can be determined from

$$A_i = \lim_{s \to p_i} (s - p_i) F(s), \qquad i = 1, \ldots, n.$$

Hence it follows, using the developed properties for transforms,

$$f(t) = \mathcal{L}^{-1}[F(s)] = A_1 e^{p_1 t} + \cdots + A_n e^{p_n t}.$$

If there is a pair of complex conjugate poles of $F(s)$ it is usually convenient to express $F(s)$ in a partial fraction expansion such that

$$F(s) = \frac{A_1 s + B_1}{s^2 + 2\xi\omega s + \omega^2} + \cdots,$$

where $s^2 + 2\xi\omega s + \omega^2 = (s - p_i)(s - p_j)$ for complex conjugate poles p_i and p_j. Then the inverse transform expression necessarily contains terms of the form

$$e^{-\xi t} \cos \omega t \qquad \text{and} \qquad e^{-\xi t} \sin \omega t.$$

This inversion procedure can be modified if the poles of $F(s)$ are *repeated*. As a simple illustration, suppose that $F(s)$ has the form

$$F(s) = \frac{N(s)}{(s - p_1)^2(s - p_2) \cdots (s - p_{n-1})},$$

where the pole p_1 is repeated twice and $p_1, \ldots, p_{n-1}$ are distinct. Then $F(s)$ can be written in the partial fraction expansion

$$F(s) = \frac{A_{11}}{(s - p_1)^2} + \frac{A_{12}}{s - p_1} + \frac{A_2}{s - p_2} + \cdots + \frac{A_{n-1}}{s - p_{n-1}}.$$

Thus the inverse transform can be shown to be of the form

$$f(t) = \mathcal{L}^{-1}\left[F(s)\right] = A_{11}te^{p_1t} + A_{12}e^{p_1t} + \cdots + A_{n-1}e^{p_{n-1}t}.$$

If some of the repeated poles of $F(s)$ occur as complex conjugate pairs the inverse transform would necessarily contain terms of the form

$$te^{-\xi t}\cos\omega t \qquad \text{and} \qquad te^{-\xi t}\sin\omega t.$$

Extensive use of these inversion procedures is made in the text, where many illustrations are given.

For reference, a brief list of some Laplace transform pairs is given in Figure B-1.

$f(t)$	$F(s)$
1	$\dfrac{1}{s}$
t	$\dfrac{1}{s^2}$
$t^n, n = 1, 2, \ldots$	$\dfrac{n!}{s^{n+1}}$
e^{-at}	$\dfrac{1}{s + a}$
te^{-at}	$\dfrac{1}{(s + a)^2}$
$\sin\omega t$	$\dfrac{\omega}{s^2 + \omega^2}$
$\cos\omega t$	$\dfrac{s}{s^2 + \omega^2}$
$e^{-at}\sin\omega t$	$\dfrac{\omega}{(s + a)^2 + \omega^2}$
$e^{-at}\cos\omega t$	$\dfrac{s + a}{(s + a)^2 + \omega^2}$

Figure B-1

Appendix C

Routh-Hurwitz Criteria

As discussed in Chapters II through IV the stability of a linear system is related to the zeros of the characteristic polynomial, which can be written in the form

$$d(s) = s^n + a_1 s^{n-1} + \cdots + a_{n-1} s + a_n$$

where $a_1, \ldots, a_n$ are real constants. The zeros of the characteristic polynomial, called the characteristic zeros, are necessarily complex valued and if the characteristic polynomial is of degree n then there are n characteristic zeros. If $\lambda_1, \ldots, \lambda_n$ denote the characteristic zeros then of course

$$d(\lambda_i) = 0, \qquad i = 1, \ldots, n$$

so that

$$d(s) = (s - \lambda_1)(s - \lambda_2) \cdots (s - \lambda_n).$$

Recall that a linear system is said to be stable if all of the characteristic zeros have real parts which are negative; thus, given a characteristic polynomial, it is important to be able to determine if, in fact, all characteristic zeros have negative real parts.

One procedure would be to factor the characteristic polynomial, as above. But this is usually possible only if $n = 2$ since in such case the quadratic formula can be used. For $n > 4$ there are no known explicit formulas for determining the characteristic zeros by factorization. Hence other procedures need to be considered. In fact, there are a number of procedures which can be used to test if all characteristic zeros have negative real parts, without explicitly determining the characteristic zeros. One such procedure is based on the Routh-Hurwitz test, which is now presented.

First, form the array with $n + 1$ rows as follows:

$$
\begin{vmatrix}
1 & a_2 & a_4 & \cdots \\
a_1 & a_3 & a_5 & \cdots \\
b_1 & b_2 & b_3 & \cdots \\
c_1 & c_2 & c_3 & \cdots \\
\cdot & \cdot & \cdot & \\
\cdot & \cdot & \cdot & \\
\cdot & \cdot & \cdot &
\end{vmatrix}
$$

using the following rules

$$
b_1 = \frac{a_1 a_2 - (1)a_3}{a_1}, \quad b_2 = \frac{a_1 a_4 - (1)a_5}{a_1}, \quad b_3 = \frac{a_1 a_6 - (1)a_7}{a_1}, \ldots,
$$

$$
c_1 = \frac{b_1 a_3 - a_1 b_2}{b_1}, \quad c_2 = \frac{b_1 a_5 - a_1 b_3}{b_1}, \quad c_3 = \frac{b_1 a_7 - a_1 b_4}{b_1}, \ldots,
$$

$$
d_1 = \frac{c_1 b_2 - b_1 c_2}{c_1}, \ldots.
$$

Note that each row, beyond the second row, is computed using the preceding two rows; a number of entries in the array are zero as each row is added.

The Routh-Hurwitz test is as follows: *All of the characteristic zeros have negative real part if*

(a) $a_1 > 0, a_2 > 0, \ldots, a_n > 0$,
(b) *all of the n + 1 numbers in the first column are positive.*

Thus the test is satisfied if both of conditions (a) and (b) are satisfied in which case the characteristic zeros all have negative real part. If either condition (a) or condition (b) is not satisfied then there must be a characteristic zero which has a real part that is zero or positive.

In order to indicate the use of the test several examples are presented.

EXAMPLE 1. Consider

$$
d(s) = s^3 + s^2 + 4s + 30.
$$

The array is easily determined to be

$$
\begin{vmatrix}
1 & 4 & 0 \\
1 & 30 & 0 \\
-26 & 0 & 0 \\
30 & 0 & 0
\end{vmatrix}
$$

Condition (a) is satisfied but condition (b) is not satisfied since the first entry in the third row is negative. Thus, from the Routh-Hurwitz criteria it follows that there is at least one characteristic zero with positive real part. In this case, the characteristic zeros can be determined to be $1 + j3$, $1 - j3$, and $-3 + j0$, which verifies the result.

EXAMPLE 2. Consider

$$d(s) = s^4 + a_1 s^3 + a_2 s^2 + a_3 s + a_4.$$

The Routh array is

$$\begin{vmatrix} 1 & a_2 & a_4 \\ a_1 & a_3 & 0 \\ b_1 & b_2 & 0 \\ c_1 & 0 & 0 \\ d_1 & 0 & 0 \end{vmatrix},$$

where

$$b_1 = \frac{a_1 a_2 - a_3}{a_1}, \quad b_2 = a_4,$$

$$c_1 = a_3 - \frac{a_1^2 a_4}{a_1 a_2 - a_3}, \quad d_1 = a_4.$$

Now condition (a) is satisfied if

$$a_1 > 0, \quad a_2 > 0, \quad a_3 > 0, \quad a_4 > 0$$

and condition (b) is satisfied if, in addition,

$$a_1 a_2 - a_3 > 0,$$

$$a_3 - \frac{a_1^2 a_4}{a_1 a_2 a_3} > 0.$$

Thus, if these indicated inequalities are satisfied, the four characteristic zeros necessarily all have negative real parts. If any of these inequalities is not satisfied there is a characteristic zero with a positive or zero real part.

Appendix D

Linear Approximations

In a number of situations throughout the presentation it is of interest to approximate a given function by another simpler function of a certain class, usually a linear function. In this section attention is given to the use of truncated Taylor series expansions as a means of obtaining such an approximation. Of course the original function must be sufficiently smooth that it has derivatives defined when required.

First consider a real function of a single variable; then for a fixed value $\bar{x}$ if the function has a first derivative at $\bar{x}$ it is natural to make the linear approximation

$$f(x) \simeq f(x) + \left(\frac{\partial f(\bar{x})}{\partial x} \right)(x - \bar{x}),$$

where the partial derivative is evaluated at the value $\bar{x}$. For values of x sufficiently close to $\bar{x}$ this approximation is quite good; if x differs substantially from $\bar{x}$ the approximation may be poor.

It is important to extend the above concept of linear approximation to the case of a function of several variables; first consider the case of two variables. Then for fixed values $\bar{x}_1$ and $\bar{x}_2$ if the function has partial derivatives at $(\bar{x}_1, \bar{x}_2)$ it is natural to make the linear approximation

$$f(x_1, x_2) \simeq f(\bar{x}_1, \bar{x}_2) + \sum_{i=1}^{2} \left(\frac{\partial f(\bar{x}_1, \bar{x}_2)}{\partial x_i} \right)(x_i - \bar{x}_i),$$

where the partial derivatives are evaluated at $\bar{x}_1, \bar{x}_2$ as indicated. If x_1 is sufficiently close to $\bar{x}_1$ and x_2 is sufficiently close to $\bar{x}_2$ then the function is reasonably well approximated by the indicated linear function.

In the general case of a function which depends on n variables $x_1, \ldots, x_n$ suppose that $f(x_1, \ldots, x_n)$ has partial derivatives defined at fixed values $\bar{x}_1, \ldots, \bar{x}_n$. Then it is natural to make the linear approximation

$$f(x_1, \ldots, x_n) \simeq f(\bar{x}_1, \ldots, \bar{x}_n) + \sum_{i=1}^{n} \left(\frac{\partial f}{\partial x_i} (\bar{x}_1, \ldots, \bar{x}_n) \right) (x_i - \bar{x}_i),$$

where the partial derivatives are evaluated at $\bar{x}_1, \ldots, \bar{x}_n$ as indicated. If each x_i is sufficiently close to $\bar{x}_i$ for $i = 1, \ldots, n$ then the function is reasonably well approximated by the indicated linear function.

These linear approximations can be thought of as the linear terms in a Taylor series expansion. However, only existence of first derivatives is required in order to make the linear approximation. There are other possible methods for obtaining linear approximations; but only the procedure just indicated is used.

A few simple examples are now indicated; in each case direct application is made of the previous formulas.

EXAMPLE 1. Consider the exponential function of a single variable

$$f(x) = e^x.$$

Let $\bar{x}$ be an arbitrary real value. For x close to $\bar{x}$ the exponential function has the linear approximation

$$e^x \simeq e^{\bar{x}} + e^{\bar{x}}(x - \bar{x}).$$

EXAMPLE 2. Consider the function of three variables

$$f(x_1, x_2, x_3) = x_1 \sin(x_2)\cos(x_3).$$

This function has partial derivatives at $x_1 = 1$, $x_2 = \pi/6$, $x_3 = \pi/6$. Formally, if x_1 is close to 1, if x_2 is close to $\pi/6$, if x_3 is close to $\pi/6$ then the function has the linear approximation

$$x_1 \sin(x_2)\cos(x_3) \simeq (1)\left(\frac{\sqrt{3}}{2} \right)\left(\frac{1}{2} \right) + \left(\frac{\sqrt{3}}{2} \right)\left(\frac{1}{2} \right)(x_1 - 1)$$

$$+ (1)\left(\frac{1}{2} \right)\left(\frac{1}{2} \right)\left(x_2 - \frac{\pi}{6} \right) + (1)\left(\frac{\sqrt{3}}{2} \right)\left(-\frac{\sqrt{3}}{2} \right)\left(x_3 - \frac{\pi}{6} \right)$$

$$\simeq \frac{\pi}{12} + \frac{\sqrt{3}}{4} x_1 + \frac{1}{4} x_2 - \frac{3}{4} x_3.$$

EXAMPLE 3. Consider the function of two variables

$$f(x_1, x_2) = \frac{1}{x_1 x_2}.$$

This function has partial derivatives at $x_1 = 1$, $x_2 = 1$. Thus if x_1 is close to 1 and x_2 is close to 1 then the function has the linear approximation

$$\frac{1}{x_1 x_2} \simeq 1 + (-1)(x_1 - 1) + (-1)(x_2 - 1)$$

$$\simeq 3 - x_1 - x_2.$$

Appendix E

Digital Computer Simulation and CSMP

The main emphasis has been on the development and use of state models for various physical processes. In many cases a detailed mathematical analysis is not possible. Consequently, a careful computer simulation, based on the state equations, may be required.

Although analog computers have been extensively used in simulation our interest is in the use of digital computers for simulation. Most procedure oriented languages, such as FORTRAN, PL/1, or ALGOL, can be used for simulation but there are advantages in using a special purpose simulation language. One such simulation language is CSMP, Continuous System Modeling Program. It is discussed in detail by Speckhart and Green; only a brief outline of the language is given here.

1. General Features of CSMP

CSMP is written primarily in FORTRAN although knowledge of FORTRAN is not required by the user of CSMP. All calculations are done in single precision arithmetic.

2. Functional Operations

There are 34 standard CSMP operations. They include: integration, differentiation, delay, limiter; step, ramp, and sine wave functions; function switches, comparators, quantizers, and some logic functions. A sample of some of the most useful operations is given in Table 1.

Table 1. Sample of CSMP-Functional Operations

Function	CSMP Statement
$Y = \int_0^t X\, dt + IC$ Integrator	$Y = \text{INTGRL}(IC, X)$ $Y(0) = IC$
$Y(t) = X(t - P) \quad t \geqslant P$ $Y(t) = 0 \quad t < P$ Dead-Time (DELAY)	$Y = \text{DELAY}\,(N, P, X)$ $P = \text{DELAY TIME}$ $N = \text{Number of points sampled}$ in interval P (integer constant)[a]
$Y = 0 \quad t < P$ $Y = 1 \quad t \geqslant P$ Step Function	$Y = \text{STEP}\,(P)$
$Y = 0 \quad t < P$ $Y = t - P \quad t \geqslant P$ Ramp Function	$Y = \text{RAMP}(P)$
$Y = 0 \quad t < P_1$ $Y = \sin(P_2(t - P_1) + P_3) \quad t \geqslant P_1$ Trigonometric Sine-Wave with Delay, Frequency and Phase Parameters	$Y = \text{SINE}\,(P_1, P_2, P_3)$ $P_1 = \text{DELAY}$ $P_2 = \text{FREQUENCY (rad/unit time)}$ $P_3 = \text{Phase shift in radians}$
$Y = X_2 \quad X_1 < 0$ $Y = X_3 \quad X_1 \geqslant 0$ Input Switch (RELAY)	$Y = \text{INSW}(X_1, X_2, X_3)$
$Y = 0 \quad X_1 < X_2$ $Y = 1 \quad X_1 \geqslant X_2$ Comparator	$Y = \text{COMPAR}\,(X_1, X_2)$
$Y = e^x$ Exponential	$Y = \text{EXP}(X)$
$Y = LN(X)$ Natural Logarithm	$Y = \text{ALOG}(X)$
$Y = \text{ARCTAN}\,(X)$ Arctangent	$Y = \text{ATAN}(X)$
$Y = \text{SIN}(X)$ Trigonometric Sine	$Y = \text{SIN}\,(X)$
$Y = \text{COS}\,(X)$ Trigonometric Cosine	$Y = \text{COS}\,(X)$
$Y = X^{1/2}$ Square Root	$Y = \text{SQRT}\,(X)$
$Y = |X|$ Absolute Value	$Y = \text{ABS}\,(X)$

[a] Unless otherwise specified, as done here, all symbolic names and constants used in defining CSMP functions must be real (i.e., floating-point) valued.

3. Format for CSMP Statements

The following general remarks pertain to the preparation of CSMP programs:

a. *Starting.* Most statements may begin in any column. A notable exception is ENDJOB which *must* begin in column 1.

b. *Ending.* As with FORTRAN, columns 73–80 are ignored by CSMP.

c. *Continuations.* Unlike FORTRAN, any statement ending with three periods (. . .) is considered to be followed by a continuation card. At most eight continuation cards are allowed. Continuation should not occur in the middle of variable names or constants.

d. *Comments.* Any card with an asterisk (*) in column 1 is considered a comment card.

4. Elements of CSMP

A system to be simulated is described to the program by a series of structure, data, and control statements.

(1) Structure statements describe the functional relationship between the variables of the problem and, taken together, define the model to be simulated.

(2) Data statements assign values to the parameters, constants, initial conditions and table entries associated with the simulation.

(3) Control statements specify options relating to the translation, execution and output phases of CSMP: run-time, integration interval and type of output.

The basic elements in the preparation of these three types of statements are numeric constants, symbolic names, operators, functions and labels.

Numeric constants are of two types: floating point constants and integer constants. Floating point constants expressed in numerical form may contain up to seven significant decimal digits, expressed either directly (e.g., $+19.3$, -121.7) or in exponential form (e.g., 14.5E3). Integer constants may contain up to ten decimal digits (e.g., 1009873). A maximum of twelve characters can be used to express a constant.

Symbolic names must start with an alphabetic character (A–Z), contain no more than six alphanumeric characters (A–Z, 0–9), and contain no embedded blanks or special characters. In addition, certain reserved words are not allowed as symbolic names. In particular, TIME is the name of the independent variable and should be used for no other purpose. Symbolic names are assumed to be real (i.e., floating point) valued, unless the user specifies otherwise.

Operators are used to indicate the basic arithmetic operations or relationship; these operators are the same as in FORTRAN:

+   addition
−   subtraction
*   multiplication
/   division
**   exponentiation
=   replacement
( ) grouping of variables and/or constants

Functions are the functional operations which perform the more complex mathematical computations, as described in Section 2 above.

Labels are the first words of CSMP data and control statements, which identify the purpose of the statements. For example, to specify the integration interval and the "finish time" for a run, one would use the label TIMER as follows:

$$\text{TIMER} \qquad \text{DELT} = 0.015, \qquad \text{FINTIM} = 250.$$

With the exception of COMMON and ENDJOB, the program examines only the first four characters of a label; hence INITIAL and INIT or PARAMETER and PARA are equally acceptable. Some of the more important labels are discussed in the following sections.

5. CSMP Source Structure

The central part of a simulation is a computer mechanism for solving the differential equations (actually the corresponding integral equations) that represent the dynamics of the model. There also may be computations to be performed before each run and/or after each run.

For example, certain parameters of a model might be considered basic; secondary parameters and initial conditions are often expressed as functions of those basic parameters; evaluation of these functions is desired just once per run. Frequently, too, one needs to perform terminal evaluation of each run. In order to satisfy such requirements, the general CSMP formulation is divided into three segments—*initial, dynamic* and *terminal,* that describe the computations to be performed before, during and after each simulation run. Each of these segments may comprise one or more *sections,* which in turn contain the *structure statements* that specify model dynamics and associated computations.

Initial Segment: Optional initial computation segment; these statements are executed only at TIME zero.

Dynamic Segment: it includes the complete description of the system dynamics, together with any other computations desired during the run. Functionally, the dynamic segment includes the ordinary differential equation representation of the system dynamics. These statements are executed repeatedly until a run termination condition is satisfied. For most models, the dynamic segment consists of a single section; for more complicated systems, it is often desirable that it be divided into several sections.

Terminal Segment: used for those computations desired after completion of each run. These statements are executed only at the completion of a run. This will often be a simple calculation based on the final value of one or more model variables, but more powerful use of this segment is readily possible. For example, one might incorporate an optimization algorithm that will modify the values of initial system parameters. If this section includes the statement CALL RERUN, CSMP will automatically be re-cycled through the simulation using the newly set parameters.

The structure segmentation of a CSMP source program is given below:

INITIAL
         ------- }Initial Segment

DYNAMIC
         ------- }Dynamic Segment

TERMINAL
         ------- }Terminal Segment

END

Note. Other translation control statements (besides INITIAL, DY-NAMIC, and TERMINAL) are as follows:

END: this statement causes a simulation run; it resets the independent variable (TIME) to zero and resets initial conditions.

CONTINUE: this statement also causes a simulation run; however, it does not reset either the independent variable (TIME) to zero or reset initial conditions.

STOP: specifies the end of use of the associated model; comes before ENDJOB; it *must* follow the last END card.

ENDJOB: this is a required statement at the end of the source program. The letters ENDJOB *must* be in columns 1–6.

6. Data Statements

Data statements are used to assign numerical values to the parameters, constants and initial conditions of the problem. These assignments may be changed between successive runs of the same model structure. The labels most frequently used for data statements are PARAMETER, INCON and CONSTANT, and these three may be used interchangeably. The format is as follows:

[Label][Variable name] = [Assigned value].

Additional assignments, if used, should be separated by a comma.

PARAMETER OMEGA = 4.0,   RO = 0.3.

A sequence of runs may be specified by placing the corresponding values to be assigned on successive runs in parentheses.

PARAMETER OMEGA = (4.0, 4.5, 5.0, 5.5).

7. Execution Control Statements

These statements are used to specify certain items relating to the actual simulation run—for example: run time, integration time and relative error. The most important features are the following:

(i) TIMER: This feature allows the user to specify the values of the following system variables by specifying some real numbers associated with the following.

PRDEL: print increment for output printing associated with the label PRINT; if both PRDEL and OUTDEL are required, the smaller is adjusted, as necessary, to be a submultiple of the larger. If a PRDEL is required but has not been specified, it is set equal either to OUTDEL (if this is required and has been specified) or to FINTIM/100.

OUTDEL: print increment for the print-plot output associated with the label PRTPLOT.

FINTIM: maximum simulation value for the independent variable; *this must be specified for each simulation.*

DELT: integration interval or step size for the independent variable TIME.

DELMIN: minimum allowable integration interval or step-size for variable-step integration methods. It is taken as $FINTIM \times 10^{-7}$ if unspecified.

(ii) FINISH: This label allows the user to specify run termination conditions in addition to FINTIM, e.g.,

FINISH ALT = 50.,   FUEL = 0.0.

This run is terminated when any of the FINISH specifications are satisfied, even if the value of the independent variable TIME is less than the FINTIM specifications.

(iii) METHOD: This label specifies the particular centralized integration routine to be used for simulation. If none is specified, RKS method is used. Names must be exactly as shown below.

ADAMS: second-order Adams integration with fixed interval;

MILNE: variable-step; fifth order; predictor-corrector Milne integration method.

RECT: rectangular integration

RKS: fourth-order Runge-Kutta with variable integration interval; Simpson's rule for error estimation.

RKSFX: fourth-order Runge-Kutta with fixed integration interval.

SIMP: Simpson's integration rule with fixed integration interval.

TRAPZ: trapezoidal integration.

As an example a possible use of the METHOD label is as follows:

METHOD MILNE

8. Output Control Statements

Output control statements are used to specify such items as the variables to be printed and/or print-plotted. The following output control statements are available:

PRINT: to specify variables to be printed at each PRDEL interval.

PRTPLOT: to specify one or more print-plot variables;

TITLE: to provide heading for PRINT output.

LABEL: to provide headings for PRTPLOT output.

RESET: to reset (nullify) conditions set by any of the preceding output control statements *except* TITLE, e.g.,

RESET LABEL.

Note that *no* continuation cards are allowed with the output control statements PRTPLOT, TITLE, or LABEL.

9. Comments on the Use of CSMP

A few cautions should be kept in mind when using CSMP, or any other simulation language. It is clear that the user must specify the particular state equations to be solved, including all parameter values and functions; the user must specify the initial state, any input function, the final time of the simulation FINTIM, and in addition the integration step size DELT and the integration method. *Such choices should be made with great care.* Mathematical analysis is particularly useful in developing enough insight to make such a choice which results in a good simulation program. Without an a priori analysis of the state model a good simulation program cannot be expected. Another caution concerns the fact that the state equations are being "solved" only approximately, based on the particular numerical integration scheme used in the simulation. For most nonpathological cases if the integration step size is sufficiently small accurate simulation can be expected. However, there are pathological cases that do occur and it may not be clear how small the step size should be in order to achieve some desired degree of accuracy. Hence the effect of numerical inaccuracies should always be examined. Again, the best benchmark to use is often based on mathematical analysis of the state equations as discussed throughout the previous chapters.

10. Sample CSMP Programs

Throughout the text a number of time responses have been indicated to illustrate some of the general features discussed in the Case Studies.

Example CSMP programs are now given for some of these case studies. All programs are particularly simple and should be easy to follow. The notation used is consistent with the individual case study examples. Typical output results are given for the specified parameter values.

```
*
*       CSMP SIMULATICN FCR CASE STUDY 3-1
*
*       SIMULATICN UNITS: X1(MGRAMS/CC), X2(MGRAMS/CC), TIME(HOURS)
*
INITIAL
        PARAMETER K1=1.0, K2=C.5
        CCNSTANT R1=0.C, R2=0.C, X10=0.02, X20=0.02
DYNAMIC
        X1DOT=-K1*X1+U
        X2DOT=K1*X1-K2*X2
        X1=INTGRL(X1C,X1DCT)
        X2=INTGRL(X20,X2DOT)
        U=R1*EXP(-R2*TIME)
TIMER DELT=0.01, OUTDEL=C.2, FINTIM=8.0
LABEL DRUG INGESTICN AND METABCLISM
PRTPLOT U,X1,X2
END
        CCNSTANT R1=0.2,R2=5.C,X1C=0.0, X20=0.0
END
STOP
ENDJOB
```

```
DRUG INGESTION AND METABCLISM

                          MINIMUM            X1     VERSUS TIME              MAXIMUM
                          6.7093E-06                                         2.0000E-02
     TIME        X1        I                                                  I
0.0           2.0000E-02   ---------------------------------------------------------+
2.0000E-01    1.6375E-02   -------------------------------------------------------- ---+
4.0000E-01    1.3406E-02   ----------------------------------------------------+
6.0000E-01    1.0976E-02   ------------------------------------------+
8.0000E-01    8.9866E-03   -----------------------------------+
1.0000E+00    7.3576E-03   ---------------------------+
1.2000E+00    6.0239E-03   --------------------+
1.4000E+00    4.9319E-03   ------------+
1.6000E+00    4.0379E-03   ----------+
1.8000E+00    3.3060E-03   --------+
2.0000E+00    2.7067E-03   ------+
2.2000E+00    2.2161E-03   -----+
2.4000E+00    1.8144E-03   ----+
2.6000E+00    1.4855E-03   ---+
2.8000E+00    1.2162E-03   ---+
3.0000E+00    9.9574E-04   --+
3.2000E+00    8.1524E-04   --+
3.4000E+00    6.6747E-04   -+
3.6000E+00    5.4647E-04   -+
3.8000E+00    4.4742E-04   -+
4.0000E+00    3.6631E-04   +
4.2000E+00    2.9991E-04   +
4.4000E+00    2.4555E-04   +
4.6000E+00    2.C104E-04   +
4.8000E+00    1.6459E-04   +
5.0000E+00    1.3476E-04   +
5.2000E+00    1.1033E-04   +
5.4000E+00    9.C332E-05   +
5.6000E+00    7.3957E-05   +
5.8000E+00    6.0551E-05   +
6.0000E+00    4.9575E-05   +
6.2000E+00    4.0589E-05   +
6.4000E+00    3.3231E-05   +
6.6000E+00    2.7207E-05   +
6.8000E+00    2.2276E-05   +
7.0000E+00    1.8238E-05   +
7.2000E+00    1.4932E-05   +
7.4000E+00    1.2225E-05   +
7.6000E+00    1.0009E-05   +
7.8000E+00    8.1947E-06   +
8.0000E+00    6.7093E-06   +
```

DRUG INGESTION AND METABOLISM

```
                           MINIMUM              X2      VERSUS TIME              MAXIMUM
                           1.0855E-C3                                           2.2497E-02
   TIME           X2       I                                                    I
0.0              2.0000E-02   ------------------------------------------------+
2.0000E-01       2.1541E-02   -------------------------------------------------+
4.0000E-01       2.2311E-02   --------------------------------------------------+
6.0000E-01       2.2497E-02   --------------------------------------------------+
8.0000E-01       2.2246E-02   -------------------------------------------------+
1.0000E+00       2.1677E-02   -----------------------------------------------+
1.2000E+00       2.0881E-02   --------------------------------------------+
1.4000E+00       1.9931E-02   -----------------------------------------+
1.6000E+00       1.8884E-02   --------------------------------------+
1.8000E+00       1.7782E-02   ----------------------------------+
2.0000E+00       1.6659E-02   -------------------------------+
2.2000E+00       1.5540E-02   ---------------------------+
2.4000E+00       1.4443E-02   ------------------------+
2.6000E+00       1.3381E-02   ---------------------+
2.8000E+00       1.2363E-02   ------------------+
3.0000E+00       1.1396E-02   ---------------+
3.2000E+00       1.0483E-02   ------------+
3.4000E+00       9.6260E-03   ----------+
3.6000E+00       8.8249E-03   ---------+
3.8000E+00       8.0792E-03   -------+
4.0000E+00       7.3875E-03   ------+
4.2000E+00       6.7475E-03   -----+
4.4000E+00       6.1571E-03   ----+
4.6000E+00       5.6134E-03   ---+
4.8000E+00       5.1139E-03   ---+
5.0000E+00       4.6556E-03   --+
5.2000E+00       4.2357E-03   -+
5.4000E+00       3.8516E-03   -+
5.6000E+00       3.5007E-03   +
5.8000E+00       3.1803E-03   +
6.0000E+00       2.8881E-03   +
6.2000E+00       2.6218E-03   +
6.4000E+00       2.3793E-03   +
6.6000E+00       2.1586E-03   +
6.8000E+00       1.9578E-03   +
7.0000E+00       1.7754E-03   +
7.2000E+00       1.6096E-03   +
7.4000E+00       1.4590E-03   +
7.6000E+00       1.3222E-03   +
7.8000E+00       1.1981E-03   +
8.0000E+00       1.0855E-03   +
```

DRUG INGESTION AND METABOLISM

```
                              MINIMUM            X1      VERSUS TIME              MAXIMUM
                              0.0                                                 2.6749E-02
     TIME             X1       I                                                  I
  0.0                0.0        +
  2.0000E-01         2.2543E-02 ----------------------------------------------+
  4.0000E-01         2.6749E-02 ------------------------------------------------+
  6.0000E-01         2.4951E-02 ----------------------------------------------+
  8.0000E-01         2.1551E-02 ---------------------------------------------+
  1.0000E+00         1.8057E-02 -------------------------------------+
  1.2000E+00         1.4936E-02 -----------------------------+
  1.4000E+00         1.2284E-02 ------------------------+
  1.6000E+00         1.0078E-02 -------------------+
  1.8000E+00         8.2588E-03 ---------------+
  2.0000E+00         6.7645E-03 ------------+
  2.2000E+00         5.5393E-03 ----------+
  2.4000E+00         4.5356E-03 --------+
  2.6000E+00         3.7136E-03 ------+
  2.8000E+00         3.0405E-03 -----+
  3.0000E+00         2.4893E-03 ----+
  3.2000E+00         2.0381E-03 ---+
  3.4000E+00         1.6687E-03 ---+
  3.6000E+00         1.3662E-03 --+
  3.8000E+00         1.1185E-03 --+
  4.0000E+00         9.1579E-04 -+
  4.2000E+00         7.4978E-04 -+
  4.4000E+00         6.1387E-04 -+
  4.6000E+00         5.0259E-04 +
  4.8000E+00         4.1149E-04 +
  5.0000E+00         3.3690E-04 +
  5.2000E+00         2.7583E-04 +
  5.4000E+00         2.2583E-04 +
  5.6000E+00         1.8489E-04 +
  5.8000E+00         1.5138E-04 +
  6.0000E+00         1.2394E-04 +
  6.2000E+00         1.0147E-04 +
  6.4000E+00         8.3078E-05 +
  6.6000E+00         6.8019E-05 +
  6.8000E+00         5.5689E-05 +
  7.0000E+00         4.5594E-05 +
  7.2000E+00         3.7330E-05 +
  7.4000E+00         3.0563E-05 +
  7.6000E+00         2.5023E-05 +
  7.8000E+00         2.0487E-05 +
  8.0000E+00         1.6773E-05 +
```

DRUG INGESTION AND METABOLISM

```
                        MINIMUM              X2    VERSUS TIME              MAXIMUM
                        0.0                                                 1.9754E-02
  TIME          X2      I                                                   I
  0.0           0.0     +
  2.0000E-01    2.6445E-03   -----+
  4.0000E-01    7.2478E-03   -----------------+
  6.0000E-01    1.1523E-02   -------------------------------+
  8.0000E-01    1.4855E-02   ----------------------------------------+
  1.0000E+00    1.7201E-02   ------------------------------------------------+
  1.2000E+00    1.8691E-02   -----------------------------------------------------+
  1.4000E+00    1.9491E-02   --------------------------------------------------------+
  1.6000E+00    1.9754E-02   ---------------------------------------------------------+
  1.8000E+00    1.9611E-02   --------------------------------------------------------+
  2.0000E+00    1.9167E-02   -------------------------------------------------------+
  2.2000E+00    1.8508E-02   -----------------------------------------------------+
  2.4000E+00    1.7701E-02   --------------------------------------------------+
  2.6000E+00    1.6798E-02   -----------------------------------------------+
  2.8000E+00    1.5839E-02   --------------------------------------------+
  3.0000E+00    1.4855E-02   ----------------------------------------+
  3.2000E+00    1.3870E-02   -------------------------------------+
  3.4000E+00    1.2901E-02   ----------------------------------+
  3.6000E+00    1.1961E-02   -------------------------------+
  3.8000E+00    1.1058E-02   ----------------------------+
  4.0000E+00    1.0198E-02   -------------------------+
  4.2000E+00    9.3855E-03   ----------------------+
  4.4000E+00    8.6214E-03   --------------------+
  4.6000E+00    7.9067E-03   ------------------+
  4.8000E+00    7.2408E-03   ----------------+
  5.0000E+00    6.6226E-03   ---------------+
  5.2000E+00    6.0504E-03   --------------+
  5.4000E+00    5.5222E-03   -------------+
  5.6000E+00    5.0355E-03   -----------+
  5.8000E+00    4.5882E-03   ----------+
  6.0000E+00    4.1776E-03   ---------+
  6.2000E+00    3.8014E-03   --------+
  6.4000E+00    3.4571E-03   -------+
  6.6000E+00    3.1425E-03   -------+
  6.8000E+00    2.8551E-03   ------+
  7.0000E+00    2.5930E-03   -----+
  7.2000E+00    2.3541E-03   -----+
  7.4000E+00    2.1365E-03   -----+
  7.6000E+00    1.9385E-03   ----+
  7.8000E+00    1.7583E-03   ----+
  8.0000E+00    1.5945E-03   ----+
```

```
*
*      CSMP SIMULATICN FCR CASE STUCY 4-2
*
*      SIMULATICN UNITS: X1(FEET), X2(FEET/SEC), X3(LB/SQ FT), TIME(SEC)
*
INITIAL
       PARAMETER W=2C0.0,G=32.2,WD=62.4,V=2.0,MU=0.5,K=0.4
       CCNSTANT F=78.0,H=80.C
       X10=H
       X20=0.0
       X30=WD*H
       D=(MU*G)/W

DYNAMIC
       X1DOT=X2
       X2DOT=-D*X2+U
       X3DOT=K*(WD*X1-X3)
       X1=INTGRL(X10,X1DOT)
       X2=INTGRL(X20,X2DOT)
       X3=INTGRL(X30,X3DOT)
       U=((W-WD*V)*G)/W-(G/W)*F
       P=X3/144.0
       Q=(X3-WD*X1)/144.0
TIMER DELT=0.01,PRDEL=0.75,CUTDEL=0.75,FINTIM=50.0
       FINISH X1=0.0
       LABEL VERTICAL ASCENT CF DIVER
       PRTPLOT X1,X2,P,Q
END
STOP
ENDJOB
```

VERTICAL ASCENT OF DIVER

```
                              MINIMUM              X1    VERSUS TIME              MAXIMUM
                             -2.1652E+00                                          8.0000E+01
    TIME           X1        I                                                    I
0.0             8.0000E+01   -------------------------------------------------------+
7.5000E-01      7.9876E+01   -------------------------------------------------------+
1.5000E+00      7.9513E+01   -------------------------------------------------------+
2.2500E+00      7.8925E+01   -------------------------------------------------------+
3.0000E+00      7.8125E+01   ------------------------------------------------------+
3.7500E+00      7.7126E+01   ------------------------------------------------------+
4.5000E+00      7.5940E+01   -----------------------------------------------------+
5.2500E+00      7.4577E+01   ----------------------------------------------------+
6.0000E+00      7.3048E+01   ---------------------------------------------------+
6.7500E+00      7.1363E+01   --------------------------------------------------+
7.5000E+00      6.9530E+01   -------------------------------------------------+
8.2500E+00      6.7558E+01   ------------------------------------------------+
9.0000E+00      6.5456E+01   ----------------------------------------------+
9.7500E+00      6.3231E+01   ---------------------------------------------+
1.0500E+01      6.0890E+01   --------------------------------------------+
1.1250E+01      5.8441E+01   ------------------------------------------+
1.2000E+01      5.5888E+01   -----------------------------------------+
1.2750E+01      5.3240E+01   ---------------------------------------+
1.3500E+01      5.0500E+01   --------------------------------------+
1.4250E+01      4.7675E+01   ------------------------------------+
1.5000E+01      4.4769E+01   ----------------------------------+
1.5750E+01      4.1788E+01   --------------------------------+
1.6500E+01      3.8735E+01   ------------------------------+
1.7250E+01      3.5615E+01   ----------------------------+
1.8000E+01      3.2431E+01   --------------------------+
1.8750E+01      2.9188E+01   ------------------------+
1.9500E+01      2.5889E+01   ----------------------+
2.0250E+01      2.2537E+01   --------------------+
2.1000E+01      1.9136E+01   ------------------+
2.1750E+01      1.5687E+01   ---------------+
2.2500E+01      1.2195E+01   --------+
2.3250E+01      8.6613E+00   ------+
2.4000E+01      5.0885E+00   ----+
2.4750E+01      1.4789E+00   --+
2.5500E+01     -2.1652E+00   +
```

VERTICAL ASCENT OF DIVER

```
                           MINIMUM                X2      VERSUS TIME              MAXIMUM
                         -4.8810E+00                                                0.0
        TIME        X2          I                                                    I
       0.0         0.0          -----------------------------------------------------+
       7.5000E-01  -3.2809E-01  ----------------------------------------------------+
       1.5000E+00  -6.3696E-01  ---------------------------------------------------+
       2.2500E+00  -9.2774E-01  -------------------------------------------------+
       3.0000E+00  -1.2015E+00  -----------------------------------------------+
       3.7500E+00  -1.4592E+00  ----------------------------------------------+
       4.5000E+00  -1.7018E+00  --------------------------------------------+
       5.2500E+00  -1.9302E+00  -------------------------------------------+
       6.0000E+00  -2.1452E+00  -----------------------------------------+
       6.7500E+00  -2.3476E+00  ---------------------------------------+
       7.5000E+00  -2.5381E+00  -------------------------------------+
       8.2500E+00  -2.7175E+00  -----------------------------------+
       9.0000E+00  -2.8864E+00  ---------------------------------+
       9.7500E+00  -3.0454E+00  -------------------------------+
       1.0500E+01  -3.1950E+00  -----------------------------+
       1.1250E+01  -3.3359E+00  ---------------------------+
       1.2000E+01  -3.4686E+00  --------------------------+
       1.2750E+01  -3.5935E+00  ------------+
       1.3500E+01  -3.7110E+00  ----------+
       1.4250E+01  -3.8217E+00  ----------+
       1.5000E+01  -3.9259E+00  --------+
       1.5750E+01  -4.0240E+00  --------+
       1.6500E+01  -4.1163E+00  -------+
       1.7250E+01  -4.2032E+00  ------+
       1.8000E+01  -4.2850E+00  ------+
       1.8750E+01  -4.3621E+00  -----+
       1.9500E+01  -4.4346E+00  ----+
       2.0250E+01  -4.5029E+00  ---+
       2.1000E+01  -4.5672E+00  ---+
       2.1750E+01  -4.6277E+00  --+
       2.2500E+01  -4.6846E+00  --+
       2.3250E+01  -4.7383E+00  -+
       2.4000E+01  -4.7887E+00  +
       2.4750E+01  -4.8363E+00  +
       2.5500E+01  -4.8810E+00  +
```

```
VERTICAL ASCENT OF DIVER

                       MINIMUM          P      VERSUS TIME            MAXIMUM
                       4.1533E+00                                     3.4667F+01
  TIME         P       I                                             I
0.0          3.4667E+01   ----------------------------------------------+
7.5000E-01   3.4662E+01   ----------------------------------------------+
1.5000E+00   3.4630E+01   ----------------------------------------------+
2.2500E+00   3.4552E+01   ----------------------------------------------+
3.0000E+00   3.4416E+01   ----------------------------------------------+
3.7500E+00   3.4213E+01   ----------------------------------------------+
4.5000E+00   3.3940E+01   ---------------------------------------------+
5.2500E+00   3.3593E+01   --------------------------------------------+
6.0000E+00   3.3174E+01   -------------------------------------------+
6.7500E+00   3.2682E+01   ------------------------------------------+
7.5000E+00   3.2120E+01   -----------------------------------------+
8.2500E+00   3.1489E+01   ---------------------------------------+
9.0000E+00   3.0792E+01   --------------------------------------+
9.7500E+00   3.0033E+01   ------------------------------------+
1.0500E+01   2.9214E+01   -----------------------------------+
1.1250E+01   2.8337E+01   ---------------------------------+
1.2000E+01   2.7407E+01   --------------------------------+
1.2750E+01   2.6425E+01   ------------------------------+
1.3500E+01   2.5395E+01   -----------------------------+
1.4250E+01   2.4319E+01   ---------------------------+
1.5000E+01   2.3200E+01   --------------------------+
1.5750E+01   2.2040E+01   ------------------------+
1.6500E+01   2.0841E+01   ----------------------+
1.7250E+01   1.9607E+01   ---------------------+
1.8000E+01   1.8338E+01   ------------------+
1.8750E+01   1.7037E+01   ----------------+
1.9500E+01   1.5705E+01   --------------+
2.0250E+01   1.4345E+01   ------------+
2.1000E+01   1.2958E+01   ----------+
2.1750E+01   1.1546E+01   --------+
2.2500E+01   1.0110E+01   -------+
2.3250E+01   8.6513E+00   ------+
2.4000E+01   7.1715E+00   ----+
2.4750E+01   5.6718E+00   --+
2.5500E+01   4.1533E+00   +
```

VERTICAL ASCENT OF DIVER

```
                        MINIMUM                 Q      VERSUS TIME              MAXIMUM
                         -0.0                                                  5.0916E+00
     TIME           Q            I                                                 I
    0.0           -0.0           +
    7.5000E-01    4.8774E-02     +
    1.5000E+00    1.7421E-01     -+
    2.2500E+00    3.5116E-01     ---+
    3.0000E+00    5.6139E-01     -----+
    3.7500E+00    7.9156E-01     -------+
    4.5000E+00    1.0322E+00     ----------+
    5.2500E+00    1.2764E+00     ------------+
    6.0000E+00    1.5196E+00     --------------+
    6.7500E+00    1.7582E+00     -----------------+
    7.5000E+00    1.9900E+00     -------------------+
    8.2500E+00    2.2136E+00     ----------------------+
    9.0000E+00    2.4280E+00     ------------------------+
    9.7500E+00    2.6328E+00     --------------------------+
    1.0500E+01    2.8277E+00     ---------------------------+
    1.1250E+01    3.0129E+00     -----------------------------+
    1.2000E+01    3.1884E+00     -------------------------------+
    1.2750E+01    3.3545E+00     ---------------------------------+
    1.3500E+01    3.5116E+00     -----------------------------------+
    1.4250E+01    3.6599E+00     ------------------------------------+
    1.5000E+01    3.7999E+00     --------------------------------------+
    1.5750E+01    3.9319E+00     ---------------------------------------+
    1.6500E+01    4.0564E+00     ----------------------------------------+
    1.7250E+01    4.1738E+00     -----------------------------------------+
    1.8000E+01    4.2844E+00     ------------------------------------------+
    1.8750E+01    4.3886E+00     -------------------------------------------+
    1.9500E+01    4.4867E+00     --------------------------------------------+
    2.0250E+01    4.5792E+00     ---------------------------------------------+
    2.1000E+01    4.6662E+00     ----------------------------------------------+
    2.1750E+01    4.7482E+00     -----------------------------------------------+
    2.2500E+01    4.8254E+00     -----------------------------------------------+
    2.3250E+01    4.8981E+00     ------------------------------------------------+
    2.4000E+01    4.9665E+00     -------------------------------------------------+
    2.4750E+01    5.0309E+00     -------------------------------------------------+
    2.5500E+01    5.0916E+00     --------------------------------------------------+
```

```
*
*       CSMP SIMULATICN FCR CASE STUCY 5-4
*
*       SIMULATICN UNITS: X(VCLTS), TIME(SEC)
*
INITIAL
       PARAMETER R1=5000.C, R2=10C00.0, C=1.0F-6, F=200.0, TAU=C.C1
       XO=E*(1.0-EXP(-TAU/(R1*C)))/(EXP(TAU/(R2*C))-EXP(-TAU/(R1*C)))
DYNAMIC
       XDCT=INSW(U-X,(U-X)/(R2*C),(U-X)/(R1*C))
       X=INTGRL(XO,XDCT)
       U=E*STEP(C.0)-E*STEP(TAL)+E*STEP(2.0*TAU)
TIMER DELT=1.0E-5, CUTDEL=7.5F-4, FINTIM=0.025
METHOD RKSFX
LABEL CIRCUIT WITH DICDE
PRTPLOT U, X
END
STOP
ENDJOB
```

CIRCUIT WITH DIODE

		MINIMUM	X	VERSUS TIME	MAXIMUM
		6.6952E+01			1.8195E+02
TIME	X	I			I
0.0	6.6952E+01	+			
7.5000E-04	8.5484E+01	--------+			
1.5000E-03	1.0143E+02	--------------+			
2.2500E-03	1.1516E+02	--------------------+			
3.0000E-03	1.2698E+02	--------------------------+			
3.7500E-03	1.3715E+02	-------------------------------+			
4.5000E-03	1.4590E+02	-----------------------------------+			
5.2500E-03	1.5344E+02	--------------------------------------+			
6.0000E-03	1.5992E+02	---+			
6.7500E-03	1.6551E+02	--+			
7.5000E-03	1.7031E+02	---+			
8.2500E-03	1.7444E+02	---+			
9.0000E-03	1.7800E+02	---+			
9.7500E-03	1.8107E+02	---+			
1.0500E-02	1.7308E+02	---+			
1.1250E-02	1.6057E+02	---+			
1.2000E-02	1.4897E+02	-----------------------------------+			
1.2750E-02	1.3821E+02	-------------------------------+			
1.3500E-02	1.2822E+02	--------------------------+			
1.4250E-02	1.1895E+02	---------------------+			
1.5000E-02	1.1036E+02	-----------------+			
1.5750E-02	1.C238E+02	--------------+			
1.6500E-02	9.4985E+01	-----------+			
1.7250E-02	8.8121E+01	---------+			
1.8000E-02	8.1754E+01	------+			
1.8750E-02	7.5846E+01	---+			
1.9500E-02	7.0365E+01	-+			
2.0250E-02	7.3475E+01	--+			
2.1000E-02	9.1099E+01	---------+			
2.1750E-02	1.0627E+02	---------------+			
2.2500E-02	1.1932E+02	---------------------+			
2.3250E-02	1.3056E+02	--------------------------+			
2.4000E-02	1.4023E+02	------------------------------+			
2.4750E-02	1.4856E+02	----------------------------------+			

```
*      CSMP SIMULATICN FCR CASE STUDY 6-4
*
*      SIMULATION UNITS: X1(FEET), X2(FEET/SEC), TIME(SEC)
*
INITIAL
     PARAMETER G=32.2, MU=C.5, W=130.0
     INCON X10=-500.0, X20=500.C
DYNAMIC
     X1DOT=X2
     X2DOT=(G/W)*U-INSW(X1,(G*MU/W)*X2,0.0)-G
     X1=INTGRL(X10,X1DCT)
     X2=INTGRL(X20,X2DCT)
     U=0.0
TIMER DELT=0.005, CUTDEL=0.5, FINTIM=75.0
FINISH X2=0.0
LABEL UNDERWATER LAUNCH CF FCCKET
PRTPLOT X1, X2
END
STOP
ENDJOB
```

UNDERWATER LAUNCH OF ROCKET

		MINIMUM	X1 VERSUS TIME	MAXIMUM
		-5.0000E+02		2.2764E+03
TIME	X1	I		I
0.0	-5.0000E+02	+		
5.0000E-01	-2.6372E+02	----+		
1.0000E+00	-5.3444E+01	--------+		
1.5000E+00	1.3680E+02	----------+		
2.0000E+00	3.1838E+02	-------------+		
2.5000E+00	4.9191E+02	---------------+		
3.0000E+00	6.5738E+02	-------------------+		
3.5000E+00	8.1481E+02	----------------------+		
4.0000E+00	9.6419E+02	-------------------------+		
4.5000E+00	1.1055E+03	-----------------------------+		
5.0000E+00	1.2388E+03	--------------------------------+		
5.5000E+00	1.3640E+03	-----------------------------------+		
6.0000E+00	1.4812E+03	--------------------------------------+		
6.5000E+00	1.5903E+03	--+		
7.0000E+00	1.6914E+03	--+		
7.5000E+00	1.7844E+03	--+		
8.0000E+00	1.8694E+03	--+		
8.5000E+00	1.9463E+03	--+		
9.0000E+00	2.0152E+03	---+		
9.5000E+00	2.0760E+03	---+		
1.0000E+01	2.1288E+03	---+		
1.0500E+01	2.1735E+03	--+		
1.1000E+01	2.2102E+03	---+		
1.1500E+01	2.2388E+03	--+		
1.2000E+01	2.2594E+03	---+		
1.2500E+01	2.2720E+03	--+		
1.3000E+01	2.2764E+03	--+		
1.3500E+01	2.2729E+03	--+		

UNDERWATER LAUNCH OF ROCKET

```
                            MINIMUM              X2    VERSUS TIME              MAXIMUM
                          -1.5195E+01                                        5.0000E+02
        TIME          X2         I                                                    I
0.0               5.0000E+02     --------------------------------------------------+
5.0000E-01        4.4586E+02     --------------------------------------------------+
1.0000E+00        3.9590E+02     -----------------------------------------+
1.5000E+00        3.7121E+02     ------------------------------------+
2.0000E+00        3.5511E+02     --------------------------------+
2.5000E+00        3.3901E+02     ------------------------------+
3.0000E+00        3.2291E+02     ----------------------------+
3.5000E+00        3.0681E+02     --------------------------+
4.0000E+00        2.9071E+02     ------------------------+
4.5000E+00        2.7461E+02     -----------------------+
5.0000E+00        2.5851E+02     ---------------------+
5.5000E+00        2.4241E+02     --------------------+
6.0000E+00        2.2631E+02     -----------------+
6.5000E+00        2.1021E+02     ----------------+
7.0000E+00        1.9411E+02     ---------------+
7.5000E+00        1.7801E+02     --------------+
8.0000E+00        1.6191E+02     ------------+
8.5000E+00        1.4581E+02     -----------+
9.0000E+00        1.2971E+02     ----------+
9.5000E+00        1.1361E+02     ---------+
1.0000E+01        9.7505E+01     --------+
1.0500E+01        8.1405E+01     -------+
1.1000E+01        6.5305E+01     ------+
1.1500E+01        4.9205E+01     -----+
1.2000E+01        3.3105E+01     ----+
1.2500E+01        1.7005E+01     ---+
1.3000E+01        9.0540E-01     -+
1.3500E+01       -1.5195E+01     +
```

```
*
*        CSMP SIMULATION FOR CASE STUDY 6-5
*
*        SIMULATION UNITS: X1(MGRAMS/CC), X2(MGRAMS/CC), TIME(HOURS)
*
INITIAL
        PARAMETER A1=0.05, A2=1.0, A3=0.5, A4=2.0,B1=1.0, B2=1.0, M1=100.0
        CONSTANT R1=50.0, K1=1.6, R2=0.0, K2=0.0
        INCON X10=100.0, X20=0.0
DYNAMIC
        X1DOT=-A1*X1*X2-A2*(X1-M1)*COMPAR(M1,X1)+B1*U1
        X2DOT=-A4*X2+A3*(X1-M1)*COMPAR(X1,M1)+B2*U2
        X1=INTGRL(X10,X1DOT)
        X2=INTGRL(X20,X2DOT)
        U1=R1*EXP(-K1*TIME)
        U2=R2*EXP(-K2*TIME)
TIMER DELT=0.002, OUTDEL=0.10, FINTIM=3.5
LABEL BLOOD SUGAR AND INSULIN LEVELS
PRTPLOT X1, X2
END
        PARAMETER A1=0.05, A2=1.0, A3=0.5, A4=2.0, B1=1.0, B2=1.0, M1=100.0
END
        CONSTANT R1=0.0, K1=0.0, R2=16.0, K2=2.0
        CONSTANT R1=0.0, K1=0.0, R2=16.0, K2=2.0
        INCON X10=120.0, X20=0.0
END
STOP
ENDJOB
```

BLOOD SUGAR AND INSULIN LEVELS

```
                              MINIMUM           X1      VERSUS TIME              MAXIMUM
                             9.9208E+01                                        1.1680E+02
       TIME        X1         I                                                  I
      0.0          1.0000E+02  --+
      1.0000E-01   1.0460E+02  ---------------+
      2.0000E-01   1.0841E+02  ---------------------------+
      3.0000E-01   1.1145E+02  -----------------------------------+
      4.0000E-01   1.1375E+02  ---------------------------------------------+
      5.0000E-01   1.1536E+02  -------------------------------------------------+
      6.0000E-01   1.1636E+02  ----------------------------------------------------+
      7.0000E-01   1.1680E+02  -----------------------------------------------------+
      8.0000E-01   1.1678E+02  -----------------------------------------------------+
      9.0000E-01   1.1637E+02  ----------------------------------------------------+
      1.0000E+00   1.1566E+02  --------------------------------------------------+
      1.1000E+00   1.1472E+02  -----------------------------------------------+
      1.2000E+00   1.1361E+02  -------------------------------------------+
      1.3000E+00   1.1239E+02  ---------------------------------------+
      1.4000E+00   1.1113E+02  ----------------------------------+
      1.5000E+00   1.0985E+02  -----------------------------+
      1.6000E+00   1.0859E+02  -------------------------+
      1.7000E+00   1.0738E+02  ---------------------+
      1.8000E+00   1.0624E+02  ------------------+
      1.9000E+00   1.0518E+02  ---------------+
      2.0000E+00   1.0420E+02  -----------+
      2.1000E+00   1.0333E+02  ---------+
      2.2000E+00   1.0254E+02  -------+
      2.3000E+00   1.0185E+02  -----+
      2.4000E+00   1.0125E+02  ----+
      2.5000E+00   1.0074E+02  ---+
      2.6000E+00   1.0031E+02  ---+
      2.7000E+00   9.9957E+01  --+
      2.8000E+00   9.9688E+01  -+
      2.9000E+00   9.9496E+01  +
      3.0000E+00   9.9365E+01  +
      3.1000E+00   9.9281E+01  +
      3.2000E+00   9.9232E+01  +
      3.3000E+00   9.9210E+01  +
      3.4000E+00   9.9208E+01  +
      3.5000E+00   9.9221E+01  +
```

BLOOD SUGAR AND INSULIN LEVELS

```
                          MINIMUM              X2      VERSUS TIME        MAXIMUM
                          0.0                                             3.3498E+00
  TIME          X2        I                                              I
  0.0           0.0       +
  1.0000E-01    1.1073E-01  -+
  2.0000E-01    3.9129E-01  -----+
  3.0000E-01    7.7544E-01  ----------+
  4.0000E-01    1.2102E+00  ------------------+
  5.0000E-01    1.6541E+00  ------------------------+
  6.0000E-01    2.0759E+00  ------------------------------+
  7.0000E-01    2.4532E+00  -----------------------------------+
  8.0000E-01    2.7711E+00  ---------------------------------------+
  9.0000E-01    3.0210E+00  -------------------------------------------+
  1.0000E+00    3.1997E+00  ----------------------------------------------+
  1.1000E+00    3.3080E+00  ------------------------------------------------+
  1.2000E+00    3.3498E+00  -------------------------------------------------+
  1.3000E+00    3.3311E+00  -------------------------------------------------+
  1.4000E+00    3.2594E+00  ------------------------------------------------+
  1.5000E+00    3.1429E+00  ----------------------------------------------+
  1.6000E+00    2.9900E+00  --------------------------------------------+
  1.7000E+00    2.8088E+00  -----------------------------------------+
  1.8000E+00    2.6072E+00  -------------------------------------+
  1.9000E+00    2.3922E+00  ----------------------------------+
  2.0000E+00    2.1701E+00  -------------------------------+
  2.1000E+00    1.9463E+00  ----------------------------+
  2.2000E+00    1.7255E+00  ------------------------+
  2.3000E+00    1.5114E+00  ---------------------+
  2.4000E+00    1.3069E+00  ------------------+
  2.5000E+00    1.1145E+00  ---------------+
  2.6000E+00    9.3570E-01  ------------+
  2.7000E+00    7.7179E-01  ----------+
  2.8000E+00    6.3189E-01  ---------+
  2.9000E+00    5.1735E-01  -------+
  3.0000E+00    4.2357E-01  ------+
  3.1000E+00    3.4679E-01  -----+
  3.2000E+00    2.8393E-01  ----+
  3.3000E+00    2.3246E-01  ---+
  3.4000E+00    1.9032E-01  --+
  3.5000E+00    1.5582E-01  --+
```

```
BLOOD SUGAR AND INSULIN LEVELS

                              MINIMUM              X1      VERSUS TIME           MAXIMUM
                              1.0CCCE+02                                         1.3113F+02
        TIME         X1       I                                                  I
        0.0          1.0000E+02   +
        1.0000E-01   1.C462E+02   -------+
        2.0000E-01   1.C856E+02   -------------+
        3.0000E-01   1.1191E+02   -------------------- +
        4.0000E-01   1.1477E+02   ------------------------ +
        5.0000E-01   1.1721E+02   ---------------------------- +
        6.0000E-01   1.1928E+02   ----------------------------- +
        7.0000E-01   1.2105E+02   ---------------------------------+
        8.0000E-01   1.2256E+02   -----------------------------------------+
        9.0000E-01   1.2385E+02   -------------------------------------------+
        1.0000E+00   1.2494E+02   ----------------------------------------------+
        1.1000E+00   1.2587E+02   ------------------------------------------------+
        1.2000E+00   1.2667E+02   -------------------------------------------------+
        1.3000E+00   1.2735E+02   ---------------------------------------------------+
        1.4000E+00   1.2792E+02   ----------------------------------------------------+
        1.5000E+00   1.2841E+02   -----------------------------------------------------+
        1.6000E+00   1.2883E+02   -------------------------------------------------------+
        1.7000E+00   1.2919E+02   -------------------------------------------------------+
        1.8000E+00   1.2950E+02   --------------------------------------------------------+
        1.9000E+00   1.2975E+02   --------------------------------------------------------+
        2.0000E+00   1.2998E+02   ---------------------------------------------------------+
        2.1000E+00   1.3016E+02   ---------------------------------------------------------+
        2.2000E+00   1.3032E+02   ----------------------------------------------------------+
        2.3000E+00   1.3046E+02   ----------------------------------------------------------+
        2.4000E+00   1.3058E+02   -----------------------------------------------------------+
        2.5000E+00   1.3068E+02   -----------------------------------------------------------+
        2.6000E+00   1.3076E+02   -----------------------------------------------------------+
        2.7000E+00   1.3083E+02   -----------------------------------------------------------+
        2.8000E+00   1.3090E+02   ------------------------------------------------------------+
        2.9000E+00   1.3095E+02   ------------------------------------------------------------+
        3.0000E+00   1.3099E+02   ------------------------------------------------------------+
        3.1000E+00   1.3103E+02   ------------------------------------------------------------+
        3.2000E+00   1.3106E+02   ------------------------------------------------------------+
        3.3000E+00   1.3109E+02   ------------------------------------------------------------+
        3.4000E+00   1.3111E+02   ------------------------------------------------------------+
        3.5000E+00   1.3113E+02   ------------------------------------------------------------+
```

BLOOD SUGAR AND INSULIN LEVELS

		MINIMUM -1.0000E+00	X2	VERSUS TIME	MAXIMUM 1.0000E+00
TIME	X2	I			I
0.0	0.0	--------------------------+			
1.0000E-01	0.0	--------------------------+			
2.0000E-01	0.0	--------------------------+			
3.0000E-01	0.0	--------------------------+			
4.0000E-01	0.0	--------------------------+			
5.0000E-01	0.0	--------------------------+			
6.0000E-01	0.0	--------------------------+			
7.0000E-01	0.0	--------------------------+			
8.0000E-01	0.0	--------------------------+			
9.0000E-01	0.0	--------------------------+			
1.0000E+00	0.0	--------------------------+			
1.1000E+00	0.0	--------------------------+			
1.2000E+00	0.0	--------------------------+			
1.3000E+00	0.0	--------------------------+			
1.4000E+00	0.0	--------------------------+			
1.5000E+00	0.0	--------------------------+			
1.6000E+00	0.0	--------------------------+			
1.7000E+00	0.0	--------------------------+			
1.8000E+00	0.0	--------------------------+			
1.9000E+00	0.0	--------------------------+			
2.0000E+00	0.0	--------------------------+			
2.1000E+00	0.0	--------------------------+			
2.2000E+00	0.0	--------------------------+			
2.3000E+00	0.0	--------------------------+			
2.4000E+00	0.0	--------------------------+			
2.5000E+00	0.0	--------------------------+			
2.6000E+00	0.0	--------------------------+			
2.7000E+00	0.0	--------------------------+			
2.8000E+00	0.0	--------------------------+			
2.9000E+00	0.0	--------------------------+			
3.0000E+00	0.0	--------------------------+			
3.1000E+00	0.0	--------------------------+			
3.2000E+00	0.0	--------------------------+			
3.3000E+00	0.0	--------------------------+			
3.4000E+00	0.0	--------------------------+			
3.5000E+00	0.0	--------------------------+			

BLOOD SUGAR AND INSULIN LEVELS

```
                            MINIMUM              X1      VERSUS TIME          MAXIMUM
                            9.9182E+01                                        1.2000E+02
     TIME          X1       I                                                 I
0.0            1.2000E+02   ----------------------------------------------------+
1.0000E-01     1.1958E+02   ---------------------------------------------------+
2.0000E-01     1.1853E+02   --------------------------------------------------+
3.0000E-01     1.1711E+02   ------------------------------------------------+
4.0000E-01     1.1550E+02   ---------------------------------------------+
5.0000E-01     1.1382E+02   ------------------------------------------+
6.0000E-01     1.1217E+02   ---------------------------------------+
7.0000E-01     1.1059E+02   ------------------------------------+
8.0000E-01     1.0912E+02   ----------------------------+
9.0000E-01     1.0778E+02   --------------------+
1.0000E+00     1.0656E+02   ----------------+
1.1000E+00     1.0547E+02   --------------+
1.2000E+00     1.0450E+02   -----------+
1.3000E+00     1.0364E+02   ---------+
1.4000E+00     1.0289E+02   --------+
1.5000E+00     1.0224E+02   -------+
1.6000E+00     1.0167E+02   -----+
1.7000E+00     1.0118E+02   ----+.
1.8000E+00     1.0075E+02   ---+
1.9000E+00     1.0038E+02   --+
2.0000E+00     1.0006E+02   --+
2.1000E+00     9.9800E+01   -+
2.2000E+00     9.9597E+01   +
2.3000E+00     9.9446E+01   +
2.4000E+00     9.9337E+01   +
2.5000E+00     9.9262E+01   +
2.6000E+00     9.9215E+01   +
2.7000E+00     9.9190E+01   +
2.8000E+00     9.9182E+01   +
2.9000E+00     9.9188E+01   +
3.0000E+00     9.9204E+01   +
3.1000E+00     9.9228E+01   +
3.2000E+00     9.9258E+01   +
3.3000E+00     9.9292E+01   +
3.4000E+00     9.9328E+01   +
3.5000E+00     9.9366E+01   +
```

BLOOD SUGAR AND INSULIN LEVELS

		MINIMUM	X2	VERSUS TIME	MAXIMUM
		0.0			2.9430E+00
TIME	X2	I			I
0.0	0.0	+			
1.0000E-01	1.3100E+00	--------------------+			
2.0000E-01	2.1450E+00	----------------------------------+			
3.0000E-01	2.6343E+00	---+			
4.0000E-01	2.8757E+00	---+			
5.0000E-01	2.9430E+00	--+			
6.0000E-01	2.8915E+00	---+			
7.0000E-01	2.7619E+00	--+			
8.0000E-01	2.5843E+00	---+			
9.0000E-01	2.3803E+00	--+			
1.0000E+00	2.1654E+00	---------------------------------------+			
1.1000E+00	1.9501E+00	----------------------------------+			
1.2000E+00	1.7418E+00	-----------------------------+			
1.3000E+00	1.5449E+00	--------------------------+			
1.4000E+00	1.3621E+00	----------------------+			
1.5000E+00	1.1949E+00	--------------------+			
1.6000E+00	1.0435E+00	-----------------+			
1.7000E+00	9.0775E-01	--------------+			
1.8000E+00	7.8692E-01	------------+			
1.9000E+00	6.8007E-01	----------+			
2.0000E+00	5.8610E-01	---------+			
2.1000E+00	5.0385E-01	--------+			
2.2000E+00	4.3216E-01	-------+			
2.3000E+00	3.6991E-01	------+			
2.4000E+00	3.1602E-01	-----+			
2.5000E+00	2.6952E-01	----+			
2.6000E+00	2.2949E-01	---+			
2.7000E+00	1.9512E-01	---+			
2.8000E+00	1.6566E-01	--+			
2.9000E+00	1.4048E-01	--+			
3.0000E+00	1.1898E-01	--+			
3.1000E+00	1.0066E-01	-+			
3.2000E+00	8.5072E-02	-+			
3.3000E+00	7.1828E-02	-+			
3.4000E+00	6.0589E-02	-+			
3.5000E+00	5.1066E-02	+			

```
     *
     *      CSMP SIMULATICN FCR CASE STUCY 7-2
     *
     *      SIMULATION UNITS: X1(CEGREES F), X2(FEET),X3(FEET/SEC), TIME(SEC)
     *
INITIAL
       PARAMETER TAU=90.0,K=10.0,G=32.2,WA=750.0,W=500.0,...
                 TA=60.0,MU=20.0
       INCON X10=180.0,X20=1CC.C,X30=0.0
       CONSTANT U=C.150
DYNAMIC
       X1DOT=-((X1-TA)/TAU)+K*U
       X2DOT =X3
       X3DOT=G*((WA/W)*(1.C-TA/X1)-1.0-(MU*X3)/W)
       X1=INTGRL(X10,X1DCT)
       X2=INTGRL(X20,X2DOT)
       X3=INTGRL(X30,X3DOT)
TIMER DELT=0.1,OUTDEL=5.0,PRCEL=5.0,FINTIM=200.0
LABEL VERTICAL MOTION OF HCT AIR BALLCCN
PRTPLOT X1,X2,X3
END
       CCNSTANT U=0.120
       FINISH X2=0.0
END
STOP
ENDJOB
```

VERTICAL MOTION OF HOT AIR BALLOON

```
                              MINIMUM              X1      VERSUS TIME              MAXIMUM
                            1.8000E+02                                           1.9337E+02
       TIME          X1     I                                                    I
       0.0           1.8000E+02    +
       5.0000E+00    1.8081E+02    ---+
       1.0000E+01    1.8158E+02    -----+
       1.5000E+01    1.8230E+02    --------+
       2.0000E+01    1.8299E+02    -----------+
       2.5000E+01    1.8364E+02    -------------+
       3.0000E+01    1.8425E+02    ---------------+
       3.5000E+01    1.8483E+02    ------------------+
       4.0000E+01    1.8538E+02    --------------------+
       4.5000E+01    1.8590E+02    ----------------------+
       5.0000E+01    1.8639E+02    -------------------------+
       5.5000E+01    1.8686E+02    ------------------------- +
       6.0000E+01    1.8730E+02    ----------------------------+
       6.5000E+01    1.8771E+02    ------------------------------+
       7.0000E+01    1.8811E+02    ----------------------------- +
       7.5000E+01    1.8848E+02    ------------------------------- +
       8.0000E+01    1.8883E+02    --------------------------------+
       8.5000E+01    1.8917E+02    ----------------------------------+
       9.0000E+01    1.8948E+02    -----------------------------------+
       9.5000E+01    1.8978E+02    ------------------------------------+
       1.0000E+02    1.9006E+02    -------------------------------------+
       1.0500E+02    1.9033E+02    ---------------------------------------+
       1.1000E+02    1.9058E+02    ----------------------------------------+
       1.1500E+02    1.9082E+02    -----------------------------------------+
       1.2000E+02    1.9105E+02    ------------------------------------------+
       1.2500E+02    1.9126E+02    -------------------------------------------+
       1.3000E+02    1.9146E+02    --------------------------------------------+
       1.3500E+02    1.9165E+02    ---------------------------------------------+
       1.4000E+02    1.9183E+02    ----------------------------------------------+
       1.4500E+02    1.9200E+02    -----------------------------------------------+
       1.5000E+02    1.9217E+02    -----------------------------------------------+
       1.5500E+02    1.9232E+02    ------------------------------------------------+
       1.6000E+02    1.9246E+02    ------------------------------------------------+
       1.6500E+02    1.9260E+02    -------------------------------------------------+
       1.7000E+02    1.9273E+02    -------------------------------------------------+
       1.7500E+02    1.9285E+02    --------------------------------------------------+
       1.8000E+02    1.9297E+02    --------------------------------------------------+
       1.8500E+02    1.9308E+02    --------------------------------------------------+
       1.9000E+02    1.9318E+02    --------------------------------------------------+
       1.9500E+02    1.9328E+02    ---------------------------------------------------+
       2.0000E+02    1.9337E+02    ---------------------------------------------------+
```

VERTICAL MOTION CF HOT AIR BALLCCN

```
                                  MINIMUM              X2     VERSUS TIME              MAXIMUM
                                1.0000E+02                                           2.1720E+02
       TIME          X2             I                                                   I
      0.0         1.0000E+02        +
      5.0000E+00   1.0010E+02       +
      1.0000E+01   1.0048E+02       +
      1.5000E+01   1.0111E+02       +
      2.0000E+01   1.0197E+02       +
      2.5000E+01   1.0307E+02       -+
      3.0000E+01   1.0438E+02       -+
      3.5000E+01   1.0589E+02       --+
      4.0000E+01   1.0759E+02       ---+
      4.5000E+01   1.0946E+02       ----+
      5.0000E+01   1.1150E+02       ----+
      5.5000E+01   1.1370E+02       -----+
      6.0000E+01   1.1604E+02       ------+
      6.5000E+01   1.1852E+02       -------+
      7.0000E+01   1.2113E+02       ---------+
      7.5000E+01   1.2387E+02       ----------+
      8.0000E+01   1.2672E+02       -----------+
      8.5000E+01   1.2968E+02       -------------+
      9.0000E+01   1.3274E+02       --------------+
      9.5000E+01   1.3590E+02       ---------------+
      1.0000E+02   1.3915E+02       ----------------+
      1.0500E+02   1.4249E+02       -----------------+
      1.1000E+02   1.4591E+02       -------------------+
      1.1500E+02   1.4941E+02       --------------------+
      1.2000E+02   1.5297E+02       ----------------------+
      1.2500E+02   1.5661E+02       ------------------------+
      1.3000E+02   1.6031E+02       -------------------------+
      1.3500E+02   1.6407E+02       ---------------------------+
      1.4000E+02   1.6789E+02       -----------------------------+
      1.4500E+02   1.7177E+02       ------------------------------+
      1.5000E+02   1.7569E+02       --------------------------------+
      1.5500E+02   1.7966E+02       ---------------------------------+
      1.6000E+02   1.8368E+02       -----------------------------------+
      1.6500E+02   1.8774E+02       -------------------------------------+
      1.7000E+02   1.9185E+02       --------------------------------------+
      1.7500E+02   1.9599E+02       ----------------------------------------+
      1.8000E+02   2.0017E+02       ------------------------------------------+
      1.8500E+02   2.0438E+02       -------------------------------------------+
      1.9000E+02   2.0862E+02       ---------------------------------------------+
      1.9500E+02   2.1290E+02       ----------------------------------------------+
      2.0000E+02   2.1720E+02       -------------------------------------------------+
```

```
VERTICAL MOTION CF HOT AIR BALLCCN

                         MINIMUM              X3      VERSUS TIME           MAXIMUM
                           0.0                                             8.6366E-01
      TIME            X3           I                                           I
    0.0              0.0           +
    5.0000E+00       4.7527E-02    --+
    1.0000E+01       1.0055E-01    -----+
    1.5000E+01       1.5018E-01    --------+
    2.0000E+01       1.9712E-01    ----------+
    2.5000E+01       2.4096E-01    -------------+
    3.0000E+01       2.8224E-01    ----------------+
    3.5000E+01       3.2096E-01    ------------------+
    4.0000E+01       3.5719E-01    ---------------------+
    4.5000E+01       3.9167E-01    -----------------------+
    5.0000E+01       4.2385E-01    -------------------------+
    5.5000E+01       4.5409E-01    ---------------------------+
    6.0000E+01       4.8279E-01    -----------------------------+
    6.5000E+01       5.0959E-01    -------------------------------+
    7.0000E+01       5.3498E-01    ---------------------------------+
    7.5000E+01       5.5876E-01    ----------------------------------+
    8.0000E+01       5.8107E-01    ------------------------------------+
    8.5000E+01       6.0250E-01    -------------------------------------+
    9.0000E+01       6.2240E-01    ---------------------------------------+
    9.5000E+01       6.4129E-01    ----------------------------------------+
    1.0000E+02       6.5904E-01    -----------------------------------------+
    1.0500E+02       6.7570E-01    ------------------------------------------+
    1.1000E+02       6.9160E-01    -------------------------------------------+
    1.1500E+02       7.0647E-01    --------------------------------------------+
    1.2000E+02       7.2035E-01    ---------------------------------------------+
    1.2500E+02       7.3385E-01    -----------------------------------------------+
    1.3000E+02       7.4642E-01    ------------------------------------------------+
    1.3500E+02       7.5832E-01    -------------------------------------------------+
    1.4000E+02       7.6939E-01    --------------------------------------------------+
    1.4500E+02       7.7999E-01    ----------------------------------------------------+
    1.5000E+02       7.8989E-01    ----------------------------------------------------+
    1.5500E+02       7.9929E-01    -----------------------------------------------------+
    1.6000E+02       8.C815E-01    ------------------------------------------------------+
    1.6500E+02       8.1638E-01    -------------------------------------------------------+
    1.7000E+02       8.2450E-01    -------------------------------------------------------+
    1.7500E+02       8.3199E-01    --------------------------------------------------------+
    1.8000E+02       8.3915E-01    --------------------------------------------------------+
    1.8500E+02       8.4573E-01    ---------------------------------------------------------+
    1.9000E+02       8.5210E-01    ---------------------------------------------------------;
    1.9500E+02       8.5815E-01    ----------------------------------------------------------+
    2.0000E+02       8.6366E-01    ----------------------------------------------------------+
```

```
VERTICAL MOTION CF HOT AIR BALLCCN

                          MINIMUM              X1     VERSUS TIME           MAXIMUM
                          1.6937E+02                                        1.8000E+02
     TIME        X1       I                                                 I
     0.0         1.8000E+02    ----------------------------------------------------+
     5.0000E+00  1.7935E+02    ---------------------------------------------------+
     1.0000E+01  1.7874E+02    -------------------------------------------------+
     1.5000E+01  1.7816E+02    -----------------------------------------------+
     2.0000E+01  1.7761E+02    ---------------------------------------------+
     2.5000E+01  1.7709E+02    -------------------------------------------+
     3.0000E+01  1.7660E+02    -----------------------------------------+
     3.5000E+01  1.7613E+02    ---------------------------------------+
     4.0000E+01  1.7569E+02    -------------------------------------+
     4.5000E+01  1.7528E+02    -----------------------------------+
     5.0000E+01  1.7488E+02    ---------------------------------+
     5.5000E+01  1.7451E+02    -------------------------------+
     6.0000E+01  1.7416E+02    -----------------------------+
     6.5000E+01  1.7383E+02    ---------------------------+
     7.0000E+01  1.7351E+02    -------------------------+
     7.5000E+01  1.7321E+02    -----------------------+
     8.0000E+01  1.7293E+02    ---------------------+
     8.5000E+01  1.7267E+02    --------------------+
     9.0000E+01  1.7241E+02    ------------------+
     9.5000E+01  1.7218E+02    ----------------+
     1.0000E+02  1.7195E+02    ---------------+
     1.0500E+02  1.7174E+02    -------------+
     1.1000E+02  1.7153E+02    -----------+
     1.1500E+02  1.7134E+02    ---------+
     1.2000E+02  1.7116E+02    --------+
     1.2500E+02  1.7099E+02    -------+
     1.3000E+02  1.7083E+02    ------+
     1.3500E+02  1.7068E+02    ------+
     1.4000E+02  1.7053E+02    -----+
     1.4500E+02  1.7040E+02    ----+
     1.5000E+02  1.7027E+02    ----+
     1.5500E+02  1.7014E+02    ---+
     1.6000E+02  1.7003E+02    ---+
     1.6500E+02  1.6992E+02    --+
     1.7000E+02  1.6981E+02    --+
     1.7500E+02  1.6972E+02    -+
     1.8000E+02  1.6962E+02    -+
     1.8500E+02  1.6954E+02    +
     1.9000E+02  1.6945E+02    +
     1.9500E+02  1.6937E+02    +
```

```
VERTICAL MOTION CF HOT AIR BALLOON

                              MINIMUM              X2      VERSUS TIME              MAXIMUM
                            -1.6526E-01                                          1.0000E+02
     TIME          X2       I                                                            I
    0.0            1.0000E+02   ----------------------------------------------------------+
    5.0000E+00     9.9916E+01   ----------------------------------------------------------+
    1.0000E+01     9.9615E+01   ----------------------------------------------------------+
    1.5000E+01     9.9102E+01   ----------------------------------------------------------+
    2.0000E+01     9.8388E+01   ----------------------------------------------------------+
    2.5000E+01     9.7482E+01   ---------------------------------------------------------+
    3.0000E+01     9.6393E+01   ---------------------------------------------------------+
    3.5000E+01     9.5131E+01   --------------------------------------------------------+
    4.0000E+01     9.3703E+01   --------------------------------------------------------+
    4.5000E+01     9.2119E+01   -------------------------------------------------------+
    5.0000E+01     9.0385E+01   ------------------------------------------------------+
    5.5000E+01     8.8509E+01   -----------------------------------------------------+
    6.0000E+01     8.6499E+01   ----------------------------------------------------+
    6.5000E+01     8.4360E+01   ---------------------------------------------------+
    7.0000E+01     8.2100E+01   --------------------------------------------------+
    7.5000E+01     7.9725E+01   -------------------------------------------------+
    8.0000E+01     7.7239E+01   ------------------------------------------------+
    8.5000E+01     7.4650E+01   ----------------------------------------------+
    9.0000E+01     7.1963E+01   ---------------------------------------------+
    9.5000E+01     6.9182E+01   --------------------------------------------+
    1.0000E+02     6.6311E+01   -------------------------------------------+
    1.0500E+02     6.3357E+01   -----------------------------------------+
    1.1000E+02     6.0323E+01   ----------------------------------------+
    1.1500E+02     5.7213E+01   --------------------------------------+
    1.2000E+02     5.4032E+01   -------------------------------------+
    1.2500E+02     5.0782E+01   -----------------------------------+
    1.3000E+02     4.7468E+01   ----------------------------------+
    1.3500E+02     4.4092E+01   --------------------------------+
    1.4000E+02     4.0659E+01   -------------------------------+
    1.4500E+02     3.7171E+01   -----------------------------+
    1.5000E+02     3.3630E+01   ---------------------------+
    1.5500E+02     3.0041E+01   -------------------------+
    1.6000E+02     2.6404E+01   -----------------------+
    1.6500E+02     2.2724E+01   ----------+
    1.7000E+02     1.9002E+01   ---------+
    1.7500E+02     1.5239E+01   -------+
    1.8000E+02     1.1440E+01   -----+
    1.8500E+02     7.6045E+00   ---+
    1.9000E+02     3.7357E+00   -+
    1.9500E+02    -1.6526E-01   +
```

VERTICAL MOTION OF HOT AIR BALLCON

```
                              MINIMUM              X3      VERSUS TIME              MAXIMUM
                             -7.8325E-01                                            0.0
      TIME         X3        I                                                      I
      0.0          0.0       ------------------------------------------------------+
      5.0000E+00   -3.8313E-02   ----------------------------------------------------+
      1.0000E+01   -8.1808E-02   --------------------------------------------------+
      1.5000E+01   -1.2300E-01   ------------------------------------------------+
      2.0000E+01   -1.6241E-01   ----------------------------------------------+
      2.5000E+01   -1.9978E-01   --------------------------------------------+
      3.0000E+01   -2.3546E-01   ------------------------------------------+
      3.5000E+01   -2.6927E-01   ----------------------------------------+
      4.0000E+01   -3.0153E-01   --------------------------------------+
      4.5000E+01   -3.3233E-01   -----------------------------------+
      5.0000E+01   -3.6121E-01   ---------------------------------+
      5.5000E+01   -3.8889E-01   -------------------------------+
      6.0000E+01   -4.1528E-01   -----------------------------+
      6.5000E+01   -4.4007E-01   ---------------------------+
      7.0000E+01   -4.6378E-01   -------------------------+
      7.5000E+01   -4.8636E-01   -----------------------+
      8.0000E+01   -5.0758E-01   --------------------+
      8.5000E+01   -5.2785E-01   ------------------+
      9.0000E+01   -5.4712E-01   ---------------+
      9.5000E+01   -5.6552E-01   -------------+
      1.0000E+02   -5.8258E-01   -----------+
      1.0500E+02   -5.9899E-01   ----------+
      1.1000E+02   -6.1459E-01   ---------+
      1.1500E+02   -6.2952E-01   --------+
      1.2000E+02   -6.4326E-01   -------+
      1.2500E+02   -6.5653E-01   -------+
      1.3000E+02   -6.6913E-01   ------+
      1.3500E+02   -6.8121E-01   -----+
      1.4000E+02   -6.9226E-01   ----+
      1.4500E+02   -7.0297E-01   ----+
      1.5000E+02   -7.1315E-01   ---+
      1.5500E+02   -7.2267E-01   --- +
      1.6000E+02   -7.3176E-01   --- +
      1.6500E+02   -7.4040E-01   --+
      1.7000E+02   -7.4866E-01   --+
      1.7500E+02   -7.5624E-01   -+
      1.8000E+02   -7.6356E-01   -+
      1.8500E+02   -7.7050E-01   +
      1.9000E+02   -7.7712E-01   +
      1.9500E+02   -7.8325E-01   +
```

References

R. L. Ackoff and M. W. Sasieni, *Fundamentals of Operations Research*, Wiley, New York, 1968.

J. A. Adams, D. G. Rogers, *Computer Aided Heat Transfer Analysis*, McGraw-Hill, New York, 1973.

J. G. Andrews and R. R. McLone, *Mathematical Modelling*, Butterworths, Boston, 1976.

P. M. Auslander, Y. Takahashi and M. J. Rabins, *Introducing Systems and Control*, McGraw-Hill, New York, 1974.

N. T. J. Bailey, *The Mathematical Theory of Infectious Diseases and Its Applications*, Hafner, 1976.

E. A. Bender, *An Introduction to Mathematical Modeling*, Wiley, New York, 1977.

D. Berlinski, *On Systems Analysis: An Essay Concerning the Limitations of Some Mathematical Methods in the Social, Political, and Biological Sciences*, MIT Press, Cambridge, Mass., 1976.

G. Birkhoff and G. C. Rota, *Ordinary Differential Equations*, Wiley, New York, 1978.

A. K. Biswas, *Systems Approach to Water Management*, McGraw-Hill, New York, 1976.

W. B. Blesser, *A Systems Approach to Biomedicine*, McGraw-Hill, New York, 1969.

W. E. Boyce and R. C. DiPrima, *Elementary Differential Equations and Boundary Value Problems*, Wiley, New York, 1969.

R. Boylestad and L. Nashelsky, *Electronic Devices and Circuit Theory*, Prentice-Hall, Englewood Cliffs, N.J., 1972.

M. Braun, *Differential Equations and their Applications*, Springer-Verlag, New York, 1975.

E. Burmeister and A. R. Dobell, *Mathematical Theories of Economic Growth*, Macmillan, London, 1970.

J. A. Cadzow, *Discrete Time Systems*, Prentice-Hall, Englewood Cliffs, N.J., 1973.

R. H. Cannon, *Dynamics of Physical Systems*, McGraw-Hill, New York, 1967.

C. Clark, *Mathematical Bioeconomics*, Wiley-Interscience, New York, 1976.

C. M. Close and D. K. Frederick, *Modeling and Analysis of Dynamic Systems*, Houghton Mifflin, Boston, 1978.

J. S. Coleman, *Introduction to Mathematical Sociology*, Macmillan, New York, 1964.

J. D'Azzo and C. Houpis, *Linear Control System Analysis and Design*, McGraw-Hill, New York, 1975.

M. L. Dertouzos, M. Athans, R. N. Spann and S. J. Mason, *Systems, Networks, and Computation: Multivariable Methods*, McGraw-Hill, New York, 1974.

C. A. Desoer and E. S. Kuh, *Basic Circuit Theory*, McGraw-Hill, New York, 1969.

S. Deutsch, *Models of the Nervous System*, Wiley, New York, 1967.

S. W. Director and R. A. Rohrer, *Introduction to Systems Theory*, McGraw-Hill, New York, 1972.

G. W. Evans, G. F. Wallace, G. L. Sutherland, *Simulation Using Digital Computers*, Prentice-Hall, Englewood Cliffs, N.J., 1967.

J. W. Forrester, *Industrial Dynamics*, MIT Press, Cambridge, Mass., 1961.

R. G. E. Franks, *Mathematical Modelling in Chemical Engineering*, Wiley, New York, 1966.

A. A. Frost and R. G. Pearson, *Kinetics and Mechanism, A Study of Homogeneous Chemical Reactions*, Wiley, New York, 1961.

G. Gandolfo, *Mathematical Methods and Models in Economic Dynamics*, American Elsevier, 1971.

H. J. Gold, *Mathematical Modeling of Biological Systems*, Wiley, New York, 1977.

M. R. Goodman, *Study Notes in System Dynamics*, MIT Press, Cambridge, Mass., 1974.

D. T. Greenwood, *Principles of Dynamics*, Prentice-Hall, Englewood Cliffs, N.J., 1965.

R. Haberman, *Mathematical Models: Mechanical Vibrations, Population Dynamics and Traffic Flow*, Prentice-Hall, Englewood Cliffs, N.J., 1977.

D. Halliday and R. Resnick, *Fundamentals of Physics*, Wiley, New York, 1974.

R. W. Hamming, *Numerical Methods for Scientists and Engineers*, McGraw-Hill, New York, 1973.

M. W. Hirsch and S. Smale, *Differential Equations, Dynamical Systems and Linear Algebra*, Academic Press, New York, 1974.

IBM System/360 Continuous System Modeling Program User's Manual, Program Number 360A-CX-16X.

S. L. S. Jacoby and J. S. Kowalik, *Mathematical Modeling with Computers*, Prentice-Hall, Englewood Cliffs, N.J., 1980.

W. Kaplan, *Operational Methods for Linear Systems*, Addison-Wesley, Reading, Mass., 1962.

W. Kaplan, *Ordinary Differential Equations*, Addison-Wesley, Reading, Mass., 1958.

D. Karnopp and R. Rosenberg, *System Dynamics: A Unified Approach*, Wiley, New York, 1975.

J. G. Kemeny and J. L. Snell, *Mathematical Models in the Social Sciences*, Ginn, 1962.

N. Keyfitz, *Introduction to the Mathematics of Population*, Addison-Wesley, Reading, Mass., 1968.

G. A. Korn and J. V. Wait, *Digital Continuous-System Simulation*, Prentice-Hall, Englewood Cliffs, N.J., 1977.

B. C. Kuo, *Automatic Control Systems*, Prentice-Hall, Englewood Cliffs, N.J., 1975.

C. A. Lane and J. G. March, *An Introduction to Models in the Social Sciences*, Harper and Row, 1975.

C. C. Lin and L. A. Segel, *Mathematics Applied to Deterministic Problems in the Natural Sciences*, Macmillan, New York, 1974.

D. Luenberger, *Introduction to Dynamic Systems: Theory, Models and Applications*, Wiley, New York, 1979.

H. R. Martens and D. R. Allen, *Introduction to System Theory*, C. E. Merrill, Columbus, Ohio, 1969.

R. M. May, *Stability and Complexity in Model Ecosystems*, Princeton University Press, Princeton, N. J., 1973.

T. M. McDonald, *Mathematical Methods for Social and Management Scientists*, Houghton-Mifflin, Boston, 1974.

J. M. Motil, *Digital System Fundamentals*, McGraw-Hill, New York, 1972.

R. Newcomb, *Concepts of Linear Systems and Controls*, Brooks/Cole, 1968.

B. Noble, *Applications of Undergraduate Mathematics in Engineering*, Macmillan, New York, 1971.

K. Ogata, *System Dynamics*, Prentice-Hall, Englewood Cliffs, N.J., 1978.

B. C. Patten, *Systems Analysis and Simulation in Ecology*, Academic Press, New York, 1971.

W. R. Perkins and J. B. Cruz, *Engineering of Dynamic Systems*, Prentice-Hall, Englewood Cliffs, N.J., 1967.

E. C. Pielou, *An Introduction to Mathematical Ecology*, Wiley, New York, 1969.

E. Polak and E. Wong, *Notes for a First Course on Linear Systems*, Van Nostrand Reinhold, New York, 1970.

N. Rashevsky, *Mathematical Biophysics*, Dover, 1960.

J. B. Reswick and C. K. Taft, *Introduction to Dynamic Systems*, Prentice-Hall, Englewood Cliffs, N.J., 1967.

D. S. Riggs, *The Mathematical Approach to Physiological Problems*, MIT Press, Cambridge, Mass., 1963.

F. S. Roberts, *Discrete Mathematical Models*, Prentice-Hall, Englewood Cliffs, N.J., 1976.

F. W. Sears and M. W. Zermansky, *University Physics*, Addison-Wesley, Cambridge, Mass., 1964.

S. Seely, *Dynamic Systems Analysis*, Reinhold, 1967.

S. Seely, *Electromechanical Energy Conversion*, McGraw-Hill, New York, 1962.

J. L. Shearer, A. T. Murphy and H. H. Richardson, *Introduction to Systems Dynamics*, Addison-Wesley, Reading, Mass., 1967.

W. Simon, *Mathematical Techniques for Biology and Medicine*, MIT Press, 1977.

J. M. Smith, *Mathematical Ideas in Biology*, Cambridge University Press, Cambridge, Mass., 1968.

J. M. Smith, *Mathematical Modeling and Digital Simulation for Engineers and Scientists*, Wiley, New York, 1977.

R. J. Smith, *Circuits, Devices and Systems*, Wiley, New York, 1976.

F. H. Speckhart and W. L. Green, *A Guide to Using CSMP*, Prentice-Hall, Englewood Cliffs, N.J., 1976.

S. M. Walas, *Reaction Kinetics for Chemical Engineers*, McGraw-Hill, New York, 1959.

D. C. White and H. H. Woodson, *Electromechanical Energy Conversion*, Wiley, New York, 1959.

B. Ziegler, *Theory of Modelling and Simulation*, Wiley, New York, 1976.

Index